D0548698

FLIGHTS OF
FANTASY

FLIGHTS OF

FANTASY

BILL GUNSTON

First Published in Great Britain in 1990 by The Hamlyn Publishing Group Limited,
a division of The Octopus Publishing Group,
Michelin House, 81 Fulham Road, London SW3 6RB

An imprint of BDD Promotional Book Company, Inc. 666 Fifth Avenue, New York, N.Y. 10103

ISBN 0792-45324-7

First Published in the United States of America in 1990 by The Mallard Press

Produced by Mandarin Offset. Printed and bound in Hong Kong

CONTENTS

INTRODUCTION

THE GREAT CREATIVE ENGINEERS, BY NATURE AND inclination, are inveterate dreamers. They are the rare ones who are capable, now and then, of those imaginative leaps that transform fantasy into reality and change for ever the way we look at the world. They operate in an area of human activity where the dividing line between success and failure is alarmingly thin. Even a simple design for, say, a soda siphon needs precise calculations and a sound technical grasp of the stresses and strains it will be subjected to in normal use.

The dream of flight, the chance to soar like a bird, to escape the ties and troubles of earthbound existence, is one of man's most ancient and persistent fantasies. This book looks at some of the visionaries who kept the dream alive down the centuries, who made the first succesful ascents above terra firma and who, in our own day, have made soaring into the empyrean as routine as catching a bus.

Looking back to the early 19th century and to the work of the pioneer aviators, it might seem as if man's progress towards conquest of the air was both systematic and inevitable. Nothing could be further from the truth. Even after the dreamers had begun to think of the flying machine as something more than a mechanical replica of a bird, and had arrived at the idea of using wings for lift and a propeller to pull the machine through the air, there were still enormous basic hurdles to overcome.

With that marvellous confidence of the true enthusiast, many of the dreamers assumed that, if they built a machine capable of flying, they could simply hop in, start the engine and take off. Stability and control, the two central and sometimes antagonistic problems of flying, hardly occurred to them; indeed, some gave little thought even to the question of how to land. Today, the idea of someone with no flying experience taking an untried aircraft into the air by himself is horrifying. Yet that, inevitably, was the problem facing the first aviators. And an equally daunting leap into the unknown was taken by test pilot 'Chuck' Yeager in 1947 when he became the first man to fly faster than sound.

No less remarkable were the Americans Orville and Wilbur Wright, who made the first-ever powered flight in an airplane, and who went about cracking the problem in as methodical and risk-free a manner as possible. Orville's epoch-making flight over the dunes at Kitty Hawk was the climax to four years of study and practice with large biplane kites and of experimental flights with a series of biplane gliders.

As is often the way, with pioneers as well as prophets, the Wright brothers met not with approbation but with indifference or disbelief in America, so that by the time the airplane finally emerged as a practical vehicle in about 1910, most of the flying, and airplane construction, was being done in Europe. Unfortunately, the Europeans were too preoccupied with preparations for war to think of airlines, and the airplane was first put to general use not in the service of man but of man's destruction. True, World War I produced airplanes that were converted, at the war's end, into vehicles capable of carrying passengers and freight and with a reasonable chance not only of reaching their destination but of getting there more or less on time. Even then, however, few people thought of the airliner as a serious rival to the train over short distances, or to the ship for travel between continents.

Contemporary postcard of Louis Bleriot and his Type XI (Mod) over the English Channel on July 25, 1909.

The 20 years between World War I and World War II were rich in dreamers. The technological advances accelerated by the 1914–1918 conflict began to open up, in the minds of aviation's visionaries, limitless prospects for both commercial and military aircraft. As had happened before, however, and would happen again in a relentless cycle of ministerial or departmental incompetence, few peacetime governments evinced the smallest interest in investing in aviation research and development, especially during the Depression years.

The fate of Frank Whittle, one of the great creative engineers of the 20th century, was depressingly typical. In 1929 he invented the turbojet engine, which would revolutionize civil and military aviation. The British government showed scant enthusiasm, however, and refused to fund the necessary research or assist construction of a prototype. Eventually Whittle was obliged to allow his patent to lapse, and it was only in 1937 that he was able to complete construction of his first engine and get it running. The Germans and the Americans both had turbojet fighters flying in 1942 – a full year ahead of the British. Government indifference to creative engineering work of potentially huge importance was shown again immediately after World War II, when the British design for the world's first supersonic aircraft was canceled.

It's safe to say that the heroic period of the aviation visionaries is over. A major advance in virtually any field of aeronautics, military or commercial, is likely to involve the investment of billions of dollars (or rubles) and is likely to be the product of an anonymous, if brilliant, team at the R & D facility of great aircraft constructors.

Indeed, it could be argued that nowadays the visionaries in the modern aircraft industry are the men who hold the purse strings. The greatest people-mover of all time is the Boeing 747, the Jumbo Jet. To build the first of these in the late 1960s Boeing had to risk capital far greater than the total net worth of the company. The advance order from Pan Am for 25 of these untried giants was certainly encouraging – but it did not even begin to pay for the development costs.

Yet Boeing's management was confident that their product was good and would eventually make a profit. They were right: it won't be long before the one thousandth 747 comes off the production line. At $150 million each, the financial side of this story at least is the stuff of dreams!

Today, advanced technology means that private owners of light planes expect their machines to be equipped with the latest avionics, and many have more sophisticated systems than those used by large airliners in the 1950s. Now, with the latest wizardry at his fingertips, the weekend pilot can navigate blind through blizzards or over hazardous mountain terrain and land blind on the right runway thousands of miles from his home base. However, even such technological achievements as these are largely regarded as commonplace.

Today's most amazing aircraft is surely the Northrop B-2, the 'stealth' advanced-technology bomber.

Fortunately for aviation, the great creative engineers remain undeterred by obstacles such as lack of funding or minimal interest. They continue to dream of new goals to conquer, inevitably succumbing to those irresistible flights of fantasy . . .

BILL GUNSTON

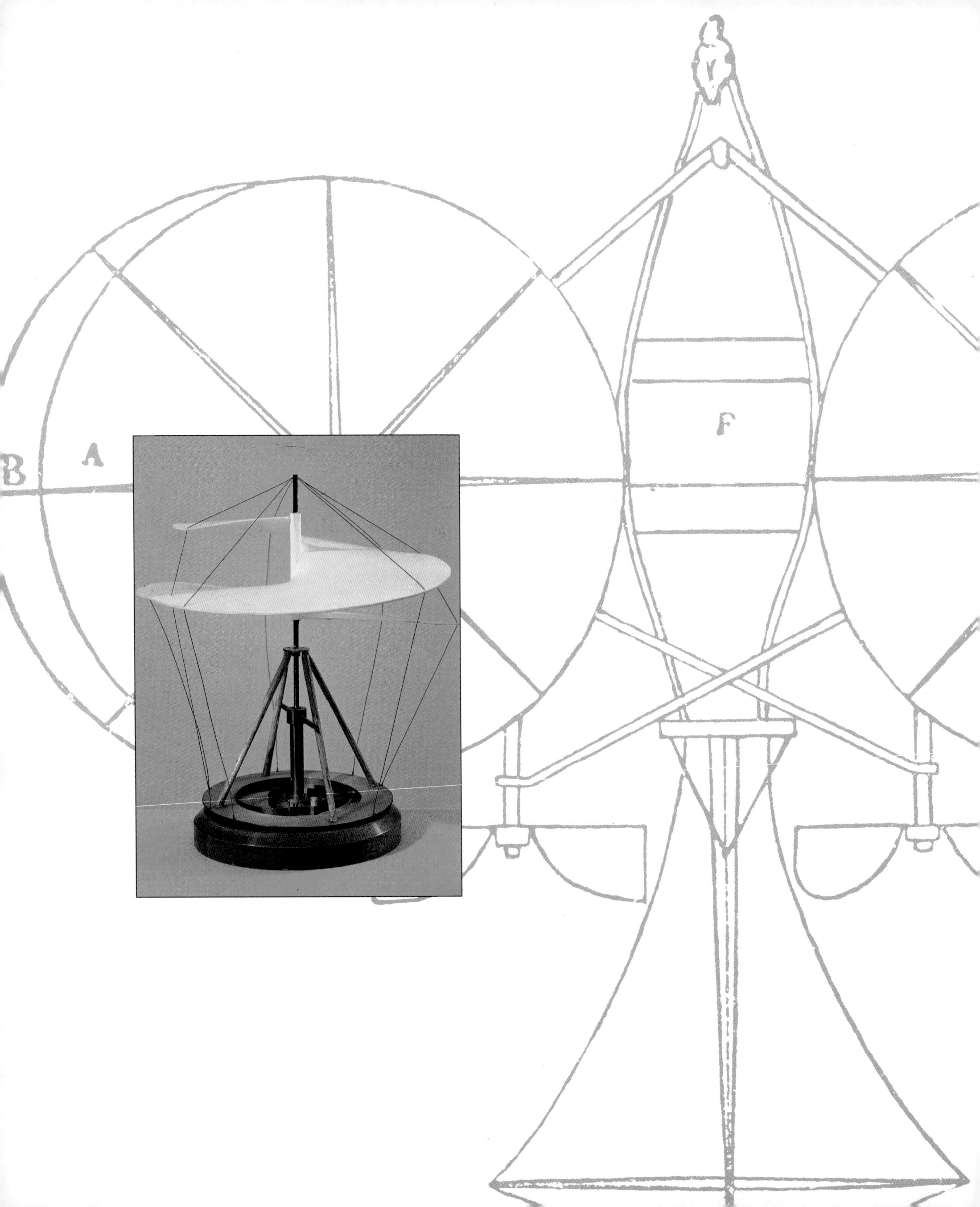
B
A
F

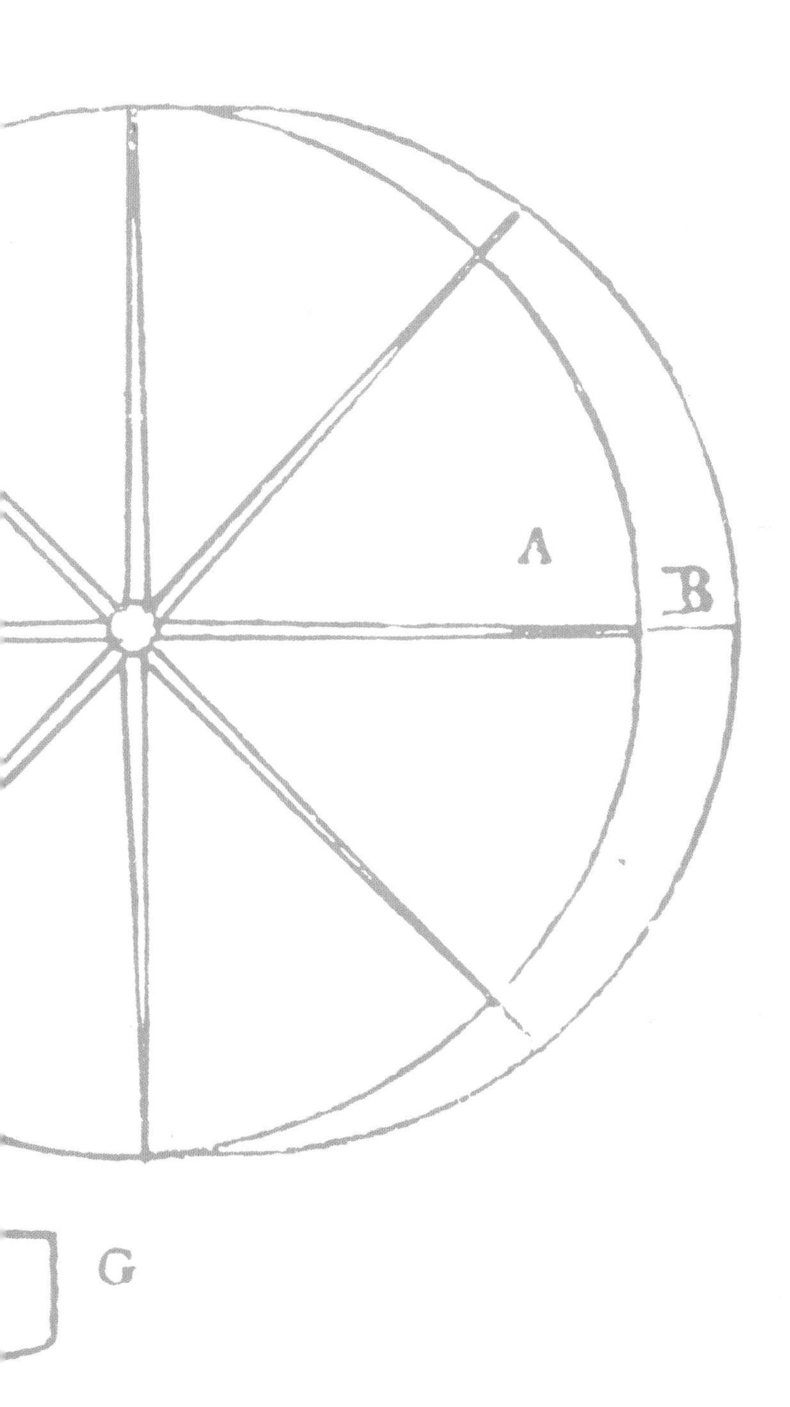

1

THE PIONEERS OF FLIGHT

Main illustration: Dating from 1843, this drawing shows one of the many flying machines proposed by Sir George Cayley. It has two biplane lifting rotors and two pusher propellers.
Inset left: Model of Leonardo da Vinci's 'helicopter'.
This page top: Lilienthal gliding over the Stöllner Hills in 1894.
This page, above: Ernst Heinkel, later a famous aircraft manufacturer, flying his home-made biplane (based on Farman designs) near Stuttgart in July 1911.

THE PIONEERS OF FLIGHT

Man has dreamed of flying for thousands of years – a dream expressed in the myths and legends of peoples of every land. For most of those long years the dream took the form of flying like some great bird. The prehistory of aviation is littered with sometimes hilarious but often tragic accounts of 'bird-men' leaping off roofs and castle turrets, flapping wing-like contrivances strapped to their arms – and plunging to the ground.

Five centuries ago the great Italian Renaissance painter, architect and engineer Leonardo da Vinci became fascinated with the technical aspects of flight and filled scores of notebooks with aeronautical speculations and drawings. His best-known design in this field was for the ornithopter, a machine with flapping wings that would enable a man to emulate the lords of the air. The several designs he produced had three things in common: they were beautifully drawn, they exhibited a visionary imagination – and they could not possibly work. Like all such bird-imitating machines, they lacked the necessary lifting power and light weight.

Although he spent hours watching birds in flight flapping their wings and soaring (gliding), Leonardo did not understand how birds actually fly. He came to the seemingly sensible conclusion that they beat their wings to push the air downwards and backwards. He thought they somehow squeezed the air under their wing-tips to make it denser, giving better support, and he was convinced they were able to soar for long periods, with their wings motionless, because they were able to use the energy in the wind. He could not know that, once a bird is airborne and well away from the Earth, there is no such thing as wind. Birds, like clouds and smoke, are carried along by the atmosphere, and the only effect of wind is to make the Earth appear to move past underneath.

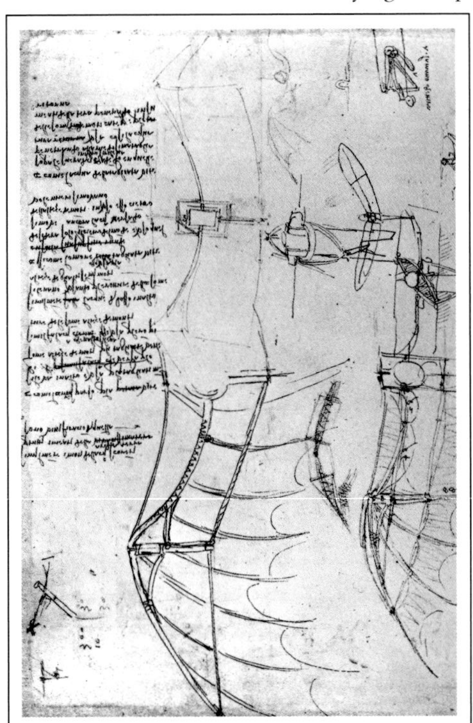

***Above*: Design sketches by Leonardo da Vinci for his proposed man-powered ornithopter.**

Although the 'bird-man' concept was a dead end, men continued to ponder other ways of getting aloft. The first successful answer was, quite literally, a lot of hot air. In the 1780s, at Annonay, south of Lyon, the French brothers Etienne and Joseph Montgolfier had a flourishing paper-making business. Like countless other people they often wondered why fragments of burning paper rose into the sky above a bonfire. Unlike most other people, however, the Montgolfiers had the kind of curiosity that leads to speculation and experiment. They came to realize that all hot air rises and they hit on the idea of trying to capture some of it in a kind of inverted paper bag. They made a big bag – the world's first balloon – and stood it with its open end above a bonfire laid in a pit. Eventually they could no longer hold the balloon down, and when they let it go it rose to a height guessed to be 1,000 feet. The date was April 25, 1783.

The next step was to make bigger balloons. On September 19, 1783 the third *Montgolfière*, some 41 feet in diameter, rose into the sky before King Louis XVI and his court at Versailles. In the basket beneath the balloon were a cockerel, a sheep and a duck. A month later a young scientist, Pilâtre de Rozier, was lifted to 85 feet in a tethered balloon. He suffered no ill effects, so on November 11, 1783 de Rozier and his friend the Marquis d'Arlandes lifted off from a garden on the outskirts of Paris and sailed in their *Montgolfière* for more than five miles, de Rozier refuelling the on-board fire with straw.

Like clouds, balloons go wherever the winds take them. And so the thoughts of would-be aviators turned to ways of navigating through the sky. Based on the analogy of ships at sea, balloons began to be equipped with sails, oars and rudders, their designers failing to realize that it is the water, not the air, that enables such equipment to steer ships at sea. And it was the prevailing westerly wind that enabled an elaborately oared and ruddered balloon to make the first manned flight across the English Channel from Dover

A modern replica of Leonardo da Vinci's ornithopter – (This lacks the essential fabric to complete the wings).

***Main illustration*: One of Lilienthal's gliders was this biplane, which he was flying throughout 1895. At the time of his death, Lilienthal was developing an engine and an elevator control.**
***Inset*: Cayley's 'Aerial Carriage' of 1843.**

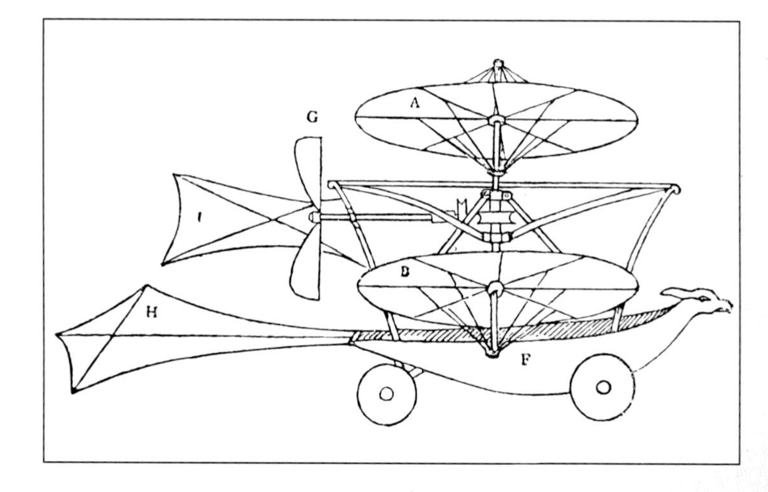

to France on January 7, 1785. It was only with the invention of the screw propeller in the mid-19th century that determining one's direction of flight became possible. Even then it needed the development of a powerful but lightweight engine to turn the possible into reality. When, from 1852, these various elements were put together, the result was the airship.

Airships have a checkered history. Today, partly owing to dreamers who could see the possibilities, they are coming back into fashion. But this book is really concerned with heavier-than-air aircraft, and especially with airplanes. Once these had been made to work they quickly became the dominant kind of aircraft, far more important than balloons and airships. Of all our flying vehicles, they most closely resemble the birds in appearance, and yet they are completely different from the birds in the way they fly. It is hard to believe that a mere century ago the successful airplane did not exist. Even when men began to fly, around the beginning of this century, their airplanes were so flimsy that often a spluttering machine just staggered into the air only to drop back to the ground and break into pieces.

The crucial difference between Leonardo's ornithopter and the airplane is that the former uses flapping wings for both lift and propulsion, whereas the latter uses one or more fixed wings solely for lift. Propulsion is provided by the engine driving a propeller or generating thrust. In fact the true ancestor of the airplane is the kite. Kites were invented so long ago that nobody knows how or where it happened, but they were quite well developed in China 2,500 years ago. Indeed, in about 500 BC Kungshu Phan described the construction of wooden kites big enough to lift men, noting that they were able to carry things into or out of a city under siege.

At first glance there might not seem to be close kinship between kites and airplanes. In fact, they are versions of the same thing: both are lifted by the flow of air over a surface held at an angle to the flow. The pull of the string on the kite stops it from being blown away, and applies a force just like the thrust of the propeller or jet engine. But of course, the very fact that a kite has to be tethered in order to work means that it cannot be used to voyage from one place to another. (With one exception, perhaps: in 1827 George Pocock, a schoolteacher, hitched two big kites to a light carriage and, with four people aboard, was drawn the 40-odd miles from Bristol to

Continued on page 16

Maxim's giant biplane at his test track at Baldwyn's Park, Kent, in 1894.

Marlborough by the force of the wind!)

Throughout the 19th century visionaries tried to make the technical leap from the concept of the static kite to that of an aerial machine that was not merely mobile, like a balloon, but directable. Until the early years of the century the miracle of bird flight had led the visionaries down a false trail by persuading them that man-made wings, like those of birds, could provide propulsion as well as lift. They can't.

One man who came to understand this and went on to lay the ground rules for the design of heavier-than-air machines was the Yorkshireman Sir George Cayley (1773–1857). We have to remember that in his day the science of aeronautics had yet to be born – indeed, there was still very little sound theoretical knowledge of the Earth's atmosphere. Cayley had to work out everything for himself, and he brought to his work a lively imagination, a naturally scientific curiosity and the ability to translate his ideas into practice in the form of model flying machines.

Like others before him, Cayley studied bird flight; unlike the others, however, he discovered how bird wings actually perform their work. He noted, for instance, that the wings do not 'row' the bird through the air by pushing backwards on the down-stroke. He saw that, on the contrary, they beat forwards on the down-stroke, and that only the feathers at the tip of each wing provide forward propulsion.

By 1804 he had worked out how and why the air exerts lift and drag on a wing and, thus, the ideal shape and curvature of its surface. He pioneered the study of the twin problems of stability and control, which showed him the need for a tail at the back of the flying machine, and the importance of having movable control surfaces in both wing and tail. All these discoveries lie at the very heart of the modern science of aeronautics.

Cayley tested his discoveries in a variety of ways. By the end of the first decade of the 19th century he had invented and built the first apparatus to be used in aerodynamic research, developed hot-air engines that worked, and made model gliders that embodied his aeronautical theories. A replica of one of these models is in the Science Museum, London. The body consists of a 5-feet-long rod. At the front end is a sliding weight which can be used to alter the glider's center of gravity; at the rear end is a simple cruciform tail unit that can be adjusted to give directional control. The wing is essentially a kite with a semicircular leading edge. The model flew perfectly, responding predictably to Cayley's subtle adjustments of the nose weight and tail unit.

Towards the end of his life Cayley built two full-size gliders, a monoplane and a triplane. On one occasion in 1853 the triplane flew across a narrow valley with Cayley's coachman on board to act as ballast. On regaining terra firma, the frightened but unscathed coachman handed in his notice, complaining that he had been employed to drive, not to fly. We have no record of his name – but he was the first man in history to make a sustained flight in a heavier-than-air machine.

Cayley predicted the coming of aerial machines 'able to transport ourselves and families, with their goods and chattels, more securely by air than by water, and with a velocity of from 20 to 100 miles per hour'. And, referring to the seemingly boundless ubiquity of the air around us, he wrote: 'An uninterrupted navigable ocean, that comes to the threshold of every man's door, ought not to be neglected as a source of human gratification and advantage.'

Cayley was, perhaps, born half a century too early. What ultimately impeded his progress was the lack of what he called a 'first mover' – a small, lightweight engine that would be reliable and powerful enough to propel his flying machines. By 1852 he had built a full-size monoplane glider equipped with pilot-managed controls, about which the historian Charles Gibbs-Smith wrote: 'This prophetic machine, if its message had been realized, could have led to practical controlled gliders in the 1850s or 1860s, which in turn might well have precipitated the powered aeroplane by the 1880s or 1890s at the latest. But Cayley's design must have appeared so visionary to his contemporaries that no-one is known to have even commented on it at the time . . .'.

By 1850, of course, the steam engine was propelling railway locomotives and other vehicles, but its power-to-weight ratio made it unsuitable for Cayley's heavier-than-air machines. It was only in the last 15 years of the century that the reliable, lightweight internal-combustion engine arrived that would enable the airplane to close the gap between dream and reality.

The advent of the glider

If Cayley provided the theoretical basis for the design of airplanes and how to control them in flight, the man who most effectively advanced the practice of gliding in the late 19th century was the German, Otto Lilienthal (1848–96), who deserves the title of the world's first pilot. Oddly enough, Lilienthal's practical work was inspired by outdated theory: his book *Bird Flight as the Basis of Aviation* was founded on the hoary conviction that airplanes should imitate the flapping flight of birds. But, more important, he insisted that the aspiring aviator must test his views – and his flying machine – in the air rather than on the drawing board.

In the 1890s he built and flew, with increasing assurance and flair, a series of what we would nowadays call hang-gliders – beautifully crafted monoplanes and biplanes, with the pilot suspended below the lower wing and controlling the glider's movements by swinging his body in order to shift the center of gravity. This is exactly the way many enthusiasts fly hang-gliders today. It sounds primitive, and perhaps it is; but it enabled Lilienthal to get the feel of how airplanes actually fly and how they can be controlled. Sadly, he crashed to his death just at the time he was beginning to study the use of control surfaces.

***Right*: The sons of Bishop Milton Wright, Orville (left) and Wilbur were as modest as they were honest. For five years (1903–9) they were thought to be either idiots or boastful liars!**

Unlike Cayley, Lilienthal left behind numerous words and photographs which exerted an immediate and widespread influence. His most successful pupil was the Scotsman, Percy Pilcher, who, after flying an excellent glider, *The Hawk*, designed and built a four-cylinder engine intended to power the first airplane. As Pilcher had already solved the problems of stability and control and knew how to fly, there is every reason to believe that by late 1900 he could have secured his place in history as the builder and flyer of the first airplane – but he crashed to his death in September 1899.

Two men who were famous long before they built airplanes were Sir Hiram Maxim and Samuel Pierpont Langley. Maxim, an American working in England, spent several years carrying out research into airplane propeller thrust and wing lift, while almost ignoring the problems of stability and control. Eventually he considered the time had come to build a full-size machine, and because he had decided on steam power, with three stokers to shovel the coal into the boiler, the result was a monster. A biplane of striking shape, with 4,000 square feet of wing area, and driven by two huge propellers almost 18 feet in diameter, it must have been quite a sight. This giant machine began careering round a specially built track in 1894. The track was meant to prevent it leaving the ground, while Maxim took measurements. On one run it was calculated that, while the weight was 8,000 lb, the lift generated was 10,000 lb, and the excess lift broke the restraining rail. But Maxim never tried to make a free flight.

Langley was a distinguished astronomer and Secretary of the Smithsonian Institution, the most eminent scientific body in the United States. In the 1890s he had made large model

Orville Wright piloting the brothers' Glider No 3 (Modified) in 1902. Wilbur is on the far wingtip; the other helper is mechanic Charlie Taylor.

airplanes which flew well, so when the U.S. War Department asked him to build a full-size piloted airplane for use in the war against Spain, he did not hesitate. By 1903 it was ready. It had front and rear monoplane wings, and two pusher propellers driven by a petrol engine designed by Langley's assistant and test pilot, Charles Manly. For reasons which were never properly explained – and which, in any event, were bad ones – Langley decided the best method of take-off would be for the machine to be catapulted from the top of a boat. Even if it had worked, this would have led to difficulties at the end of the flight because the machine had no wheels or skids. But it never got that far: on both its attempted flights the machine dived straight into the water (on the second occasion after total structural failure as soon as the catapult was fired).

By this time it was 'common knowledge' that man would never fly, except in balloons. After Langley's abject failure anyone who tried to build a flying machine was immediately dubbed a lunatic. The *New York Times*, reflecting the opinion of its readers, proclaimed 'the flying machine that will really fly might be evolved ... in from one to ten million years.' It actually was to take only another nine days.

On December 14, 1903 Wilbur Wright won the toss of a coin and attempted the first take-off in the Wright powered *Flyer*. He flew only a short distance before slumping into the sand, causing slight damage. Three days later after repairs, it was brother Orville's turn. He got away cleanly and flew 120 feet in 12 seconds, into a strong wind. A little later he covered 852 feet in 59 seconds, again into a strong wind. This was the

Continued on page 23

The famous photograph showing the start of the first succesful airplane flight, on December 17, 1903. Wilbur has just let go of the wingtip. The bench and footmarks in the sand show where the *Flyer* started its takeoff.

first time an aeroplane had flown under full control of the pilot at all times and landed at a place not lower than the point at which it took off.

Almost the whole of modern aviation dates from this moment, yet hardly anybody knew about it at the time, and even five years later the Wrights were still either unknown or regarded as knaves who had claimed to do something that everyone knew to be impossible.

The brothers had a cycle shop in Dayton, Ohio. Years after they had begun flying near Dayton the editor of the local paper wrote: 'They seem like decent enough young men, yet there they were, neglecting their business to waste their time day after day on that ridiculous flying machine. I had an idea that it must worry their father.' Certainly, no-one in Dayton, and few people anywhere else in the world, had the smallest understanding of what the brothers had achieved and how it would revolutionize human life.

Like many great pioneers in every field, they made it all seem easy. They studied everything they could find about flying machines, and soon concluded that most of it was either inaccurate or complete rubbish. So they constructed a wind tunnel in which to test everything, and built a succession of gliders in order to study the problems of stability and control: and to learn how to fly. They were the first to think about lateral control. In those days the inner tubes of bicycle tires came in long, narrow boxes, and when these were open at the end, they were easy to twist. This gave the brothers the idea of using biplane wings which were deliberately made flexible so that they could be warped (twisted) to make the machine roll to left or right. They also decided to put an elevator at the front of the plane and a rudder at the back.

Most other experimenters tried to make flying machines self-stabilizing so that they would not need to be actively 'flown'. The Wrights deliberately made all their machines so that the pilot had to control them the whole time, making them climb or descend, and turn or roll to left or right. They had so much gliding experience that, by the time they built their first powered machine in 1903, their only doubts concerned the engine and propeller.

However, their mechanic Charlie Taylor masterminded the design and construction of a four-cylinder petrol engine developing about 12 horsepower, while the brothers designed and built by far the most efficient air propellers ever made up to that time. They chose to use two pusher propellers, driven by chains and geared to turn much more slowly than the engine.

***Left*: Orville Wright on his first flight at Fort Myer, near Washington DC, on September 3, 1908. Tragically, an in-flight failure just two weeks later resulted in a crash which killed Lt T.E. Selfridge, the first person killed in an airplane.**

Their 1903 flights had been over the windswept and mosquito-infested sand dunes of Kitty Hawk, on the coast of North Carolina. The following year they did their flying about eight miles east of their home town. Their best flight, in November, lasted more than five minutes, but the local paper did not think it worth reporting. In 1905 the Wrights built *Flyer III*, with many small changes and a better engine, and were soon making flights of over half an hour. By now they were complete masters of all the basic flying movements. For the first time, humans could get aboard an airplane, start the engine, and fly anywhere they wanted to go.

What makes their achievement doubly remarkable is that the brothers were isolated, whereas in Europe – and especially in France – there were dozens of enthusiasts who daily gathered to talk about fast cars, speedboats, airships and, increasingly, flying machines. In 1905 the Voisin brothers constructed a biplane glider which was towed along above the Seine by a speedboat, and they gradually built up a flourishing business making powered biplanes for customers. In November 1906 the Brazilian, Alberto Santos-Dumont, an airship pioneer, flew his strange tail-first monoplane in a Paris park for about 20 seconds, covering 720 feet. In the same year several other experimenters either just managed or just failed to fly. One who twice got his wheels off the ground in a brief hop was Trajan Vuia, a Romanian member of the Paris aviation fraternity. What made his airplane important was that it had the layout that posterity decided was the best: a tractor (propeller-in-front) monoplane, with all the tail surfaces at the back. This inspired Leon Levavasseur to build his beautiful and successful monoplanes called Antoinettes, with engines of the same name. It also led to a series of monoplanes created by Louis Blériot.

By 1909 scores of European pilots had flown, most of them in the Paris area. The Wrights had at last achieved full recognition as the first and greatest aviators; they had taken many people aloft as passengers, had given brilliant demonstrations in France, and had even sold an airplane to the U.S. Army Signal Corps. The brothers had tried to interest the U.S. War Department as early as January 1905, but had met a point-blank refusal on the grounds that Washington did not wish to finance such a crazy idea as a flying machine. The brothers wrote back, emphasizing that they did not want financing but were offering a finished product. They received the same reply as before, and after two years of trying to get the facts across, the brothers gave up. Even so, the U.S. Army would still be the first military service to buy an airplane, so initiating the idea of an 'air force'. Indeed, by 1910 some high-spirited young American officers amused themselves by firing rifles and dropping home-made bombs from aircraft. They would sit on the leading edge of the lower wing, without any seat harness or protection against the slipstream.

Alberto Santos-Dumont's first airplane which was completed in September 1906. It had the tail at the front and the propeller at the back. His best flight covered 722 ft, taking about 21 seconds on November 12, 1906.

Left: **Louis Blériot still in his flying helmet, after landing at 5.17 a.m. on a steep bank near Dover Castle, England.**
Below: **Louis Blériot's pioneer flight from France to England on July 25, 1909 made headline news.**

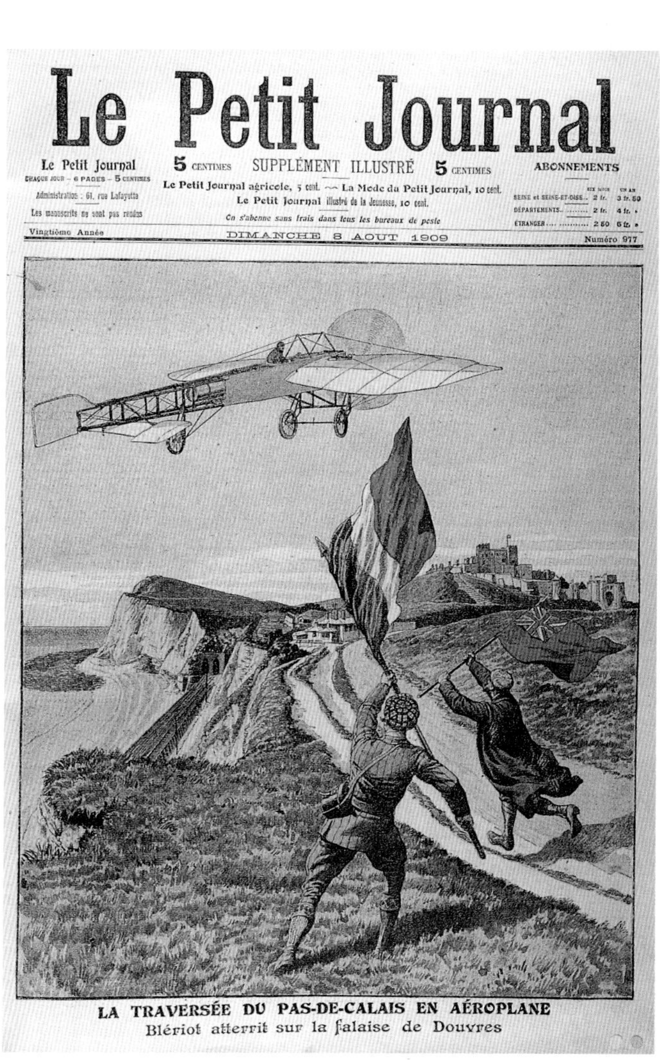

Le Petit Journal

Le Petit Journal
CHAQUE JOUR – 6 PAGES – 5 CENTIMES
Administration : 61, rue Lafayette
Les manuscrits ne sont pas rendus

5 CENTIMES SUPPLÉMENT ILLUSTRÉ 5 CENTIMES
Le Petit Journal agricole, 5 cent. ~~ La Mode du Petit Journal, 10 cent.
Le Petit Journal illustré de la Jeunesse, 10 cent.
On s'abonne sans frais dans tous les bureaux de poste

ABONNEMENTS

	SIX MOIS	UN AN
SEINE et SEINE-ET-OISE..	2 fr.	3 fr. 50
DÉPARTEMENTS..........	2 fr.	4 fr. »
ÉTRANGER..............	2 50	5 fr. »

Vingtième Année DIMANCHE 8 AOUT 1909 Numéro 977

LA TRAVERSÉE DU PAS-DE-CALAIS EN AÉROPLANE
Blériot atterrit sur la falaise de Douvres

2

WINGS FOR WAR

Left: Rockets being rippled (fired at very brief set intervals) from a Harrier GR.3 'jump jet' of Britain's Royal Air Force. Each rocket has a caliber of 2.68 in.

WINGS FOR WAR

Over the past 50 years or so, war or the threat of war has proved to be the most powerful of all stimuli to scientific invention. Nothing, it seems, so effectively attracts the investment of national funds as the latest breakthrough in the technology of death.

It was not always like that. In the period leading up to the outbreak of World War I in August 1914, all the European powers had invested massively, if not always wisely, in naval construction. But few politicians or armed-forces commanders seemed able or willing to envisage a role for the airplane as a combat vehicle. On the contrary, they seemed determined to give voice to opinions on the subject that, within a few months of war breaking out, would make them look ridiculous. The British Admiralty, for instance, let it be known that 'Their Lordships foresee no practical application for flying machines in naval service.' The British Minister of War said: 'We do not consider that aeroplanes will be of any possible use for war purposes.' The Chief of the Imperial General Staff – with the same enviable confidence that enabled him to know that the tank would never replace the cavalry – pointed out that 'We have done very well without aeroplanes so far; we can do without them to-day.' Other countries were quick to deny Britain a monopoly in such wisdom. Marshal Foch, the great French master of strategy, declared: 'Aviation is good sport, but for the Army it is useless.' And an official Appreciation by the German General Staff in September 1914 insisted that 'Experience has shown that real combat in the air, such as journalists and romancers have described, should be considered a myth.'

Given that it was these same people and others like them who considered trench warfare to be an acceptable way to conduct hostilities between nations, these views are not perhaps surprising. What is surprising is that they made no acknowledgment, let alone appraisal, of the fact that European aircraft had gone to war almost three years before the guns were first heard on the Western Front. In October 1911 a shooting war had broken out between Italy and Turkey in what today is Libya, and various airplanes were used by the Italian forces, first for reconnaissance and later to drop small bombs.

Turkey was again on the receiving end of bombs in a war in the Balkans in 1912–13, but the bombs were little more powerful than grenades. One pilot would fly with bombs suspended from wires looped over his feet; when he thought he was in the right position over enemy troops he had merely to kick off the wires and the bombs would drop. But one or two visionaries could imagine much bigger airplanes dropping much bigger bombs. Airships, too, could carry bombs, and when World War I began the German Zeppelin and Schutte-Lanz airships were regarded by the Allies as a far greater threat than any airplane. For some years these airships made raids against England, but the British defenses got better all the time and by 1918 the airships were being shot down at a rate that forced their operations to be abandoned. The German air force turned instead to large bombers. But the pioneers of such aircraft were the Italians and the Russians.

Perhaps the greatest visionary of airpower at that time was the commander of the Italian Battaglione Aviatori, Colonel Giulio Douhet. In 1913 he wrote a classic book on strategic airpower – then a wholly new concept. He predicted almost all the effects of the later heavy bombing of enemy heartlands, and he influenced the thinking (and career) of almost every subsequent military leader. Whereas many generals and admirals had poked fun at the little bombs dropped on Turkish troops in 1912, Douhet could see that the bombs were small only because they were carried by little airplanes that could hardly stagger into the air if their pilots happened to be heavier than average. For airpower to become a reality, bigger and more powerful airplanes were needed and in late 1912 he suggested this to the industrialist Count Gianni Caproni. The first big Caproni bomber flew in 1913. It looked like a very large biplane, with the tail carried on two slender booms and with a single pusher propeller at the rear of the central nacelle. This nacelle housed three Gnome engines in a row, the rearmost one driving the pusher propeller and the others driving a differential transmission to left and right tractor propellers.

This aircraft was the first of hundreds of very capable bombers which were used by Italy and other Allied countries, including Britain, and were also made under license in France and the United States. Almost all had the same arrangement of

Above right: **A Sikorsky IM four-engined bomber, operating on skis. These were the first large bombers to go into production.** ***Right:*** **The Fokker E.III was one of the series of monoplanes which devastated Allied airplanes in 1915–1916.**

a central pusher engine and two tractor propellers, but the troublesome transmission was later abandoned and the tractor engines were mounted on the front of the two tail booms, driving their propellers directly.

A Russian Pioneer

The only other country where large bombers could be seen at the start of World War I was Russia. Here the young designer Igor Sikorsky (1889–1972) – one of the greatest pioneers in the history of aviation, later famous for his Pan Am flying boats and for launching the world's helicopter industry – built his first really large biplane at the Russo-Baltic Wagon Works and flew it on March 2, 1913. By this date over 1,000 airplanes of some 500 different types had been built since the breakthrough at Kitty Hawk, and most of them had managed to fly. But not one of them weighed more than 2,000lb or had a wing span greater than 50 feet. Thus we can see how bold was young Sikorsky in making his machine weigh nearly 9,000lb loaded, with a wing span of 88ft 7in. He fitted two six-cylinder Argus engines, very like large editions of those used in cars of the day; each could produce 100 horsepower, but this was hardly enough for so big an airplane. The fuselage was amazingly slender, but at the front was a big cabin with glass windows all round and with double glass doors in front of the two pilots to enable people to pass through to the open balcony in the nose. Behind the pilots double doors opened into a cabin nearly 20 feet long with small chairs, a sofa, a table, a wardrobe for coats and even a toilet. Nothing like this had even been dreamed of before.

The big airplane was called the *Grand*, and Tsar Nicholas II came to see it. Sikorsky soon decided it needed more power, and he added two more engines. He was afraid of the asymmetric thrust that would result from failure of an engine, so he added the extra power directly behind the original engines, driving pusher propellers behind the wings. Flying was resumed on May 10, 1913 and Sikorsky was delighted to find that the modified machine could be flown with both engines on one side throttled back. He knew, however, that pusher propellers operating in the slipstream from tractor propellers in front are inherently inefficient, so he took a deep breath and modified the aircraft yet again, fitting four tractor engines along the front of the wing. To be on the safe side he added two more rudders, making four in all.

The finally rebuilt aircraft was renamed *Russki Vityaz* (Russian Knight). It can be regarded as the prototype of all the tens of thousands of four-engined aircraft that have followed it. Its loaded weight had gone up to 9,259lb, and it was a most impressive machine, even though its performance remained modest, with a ceiling of about 2,000 feet and a maximum speed of 56mph. In a sensible world it would have led to a succession of ever-better transports for passengers and cargo; but in 1913 most European countries were preparing to fight their neighbors. So the next big Sikorsky was a military airplane, designed to fly long reconnaissance missions and to

***Far left*: In 1918 the Albatros D.V was one of the chief German fighters.**
***Above*: One of the very few genuine combat photographs from World War I, showing an Albatros C.III two-seater.**

carry bombs to be dropped on any worthwhile target that might be seen. It was bigger than its predecessor, with a wing span of 105 feet. The fuselage was redesigned with a completely transparent, yet enclosed nose, and the rest enlarged to provide room for a full-length cabin. An unexpected feature was a rear 'promenade deck' on top of the plywood-covered rear fuselage, with access by a ladder and a handrail all round.

This second machine was called the *Ilya Mouromets*, after a legendary hero who had defended Kiev 1,000 years ago. It was the first of 80 aircraft, all with the same 'IM' designation, even though hardly any two were alike (on several occasions, in fact, the whole design was altered). Whereas the *Grand* had begun life with a total of 200 horsepower, the final batch of IM bombers had four Renault engines of 220 horsepower each. Loaded weight went up to 16,446lb and maximum speed to 85mph. While their chief value lay in their ability to undertake long-range photographic missions, the IMs also dropped thousands of bombs and caused such alarm to German troops that the German high command even issued a proclamation stating that the reports of giant Russian aircraft were merely rumors, and that 'such aircraft did not exist'! Although they looked flimsy, the IMs were immensely durable, and only two were shot down – one of them after its own gunners had shot down three German fighters.

In contrast, Britain and France had only very small airplanes in 1914. When the first squadrons of the Royal Flying Corps flew to France at the start of World War I in

The Sopwith Camel is now known to have shot down at least 3,000 aircraft in World War I, far more than any other fighter at that time.
***Inset*: Some of its victims were Fokker Dr. I triplanes.**

August 1914 – suffering several crashes and numerous forced landings on the way – one pilot dared to ask his commander the question many were pondering: 'Sir, what do we do if we meet a Zeppelin coming the other way?' Suddenly, air warfare, a concept ridiculed by the generals for so long, appeared to be inevitable. The official answer was that one should attempt to 'interfere' with the progress of the enemy – but how could an unarmed airplane do this except by a suicidal head-on collision?

Fighter Armaments

Pioneer aviators in several countries had tried to imagine aerial warfare, and had experimented by firing with revolvers, carbines, rifles and various kinds of machine gun while aloft. The most common airplane in the RFC in August 1914 was the B.E.2, a two-seater biplane, which had not been designed to carry any armament. Those that flew to France in the first weeks of the war often carried an observer in the front cockpit armed with a carbine which could be loaded with single rounds from a bandolier draped over the shoulder; but he had a very limited arc of fire and had to be careful not to shoot through a vital bracing wire or the propeller. Some airplanes had been specially designed with the engine arranged as a pusher, the tail being carried on booms spaced widely enough apart to leave room for the propeller behind the wings. Thus a gunner in the nose could fire a machine gun without worrying about hitting his own propeller, wings or fuselage. One such machine, the D.H.2, was an agile single-seater, with its gun fixed to fire ahead and aimed by the pilot aiming the whole aircraft.

In general these pusher airplanes were clumsy, and slow in comparison with those with the engine at the front. Several engineers had invented mechanisms for controlling the fire of a machine gun so that it could fire ahead safely, the bullets always passing between the blades of the revolving propeller. In France a famous pre-war designer, Raymond Saulnier (he had done most of the design work for Blériot's Channel-crossing airplane in 1909) had done a lot of work on such 'interrupter gears', in the course of which faulty ammunition had sometimes resulted in bullets hitting the propeller. To protect the blades he produced sets of crude steel deflector wedges. Two of these deflectors were introduced into the

Continued on page 40

D8311 was one of a batch of 50 Handley Page O/400 bombers built by the British Caudron Co. It had the usual Rolls-Royce Eagle engines, but some batches had Sunbeam engines while those built in the USA had the Liberty 12. All the engines were about 350 horsepower.

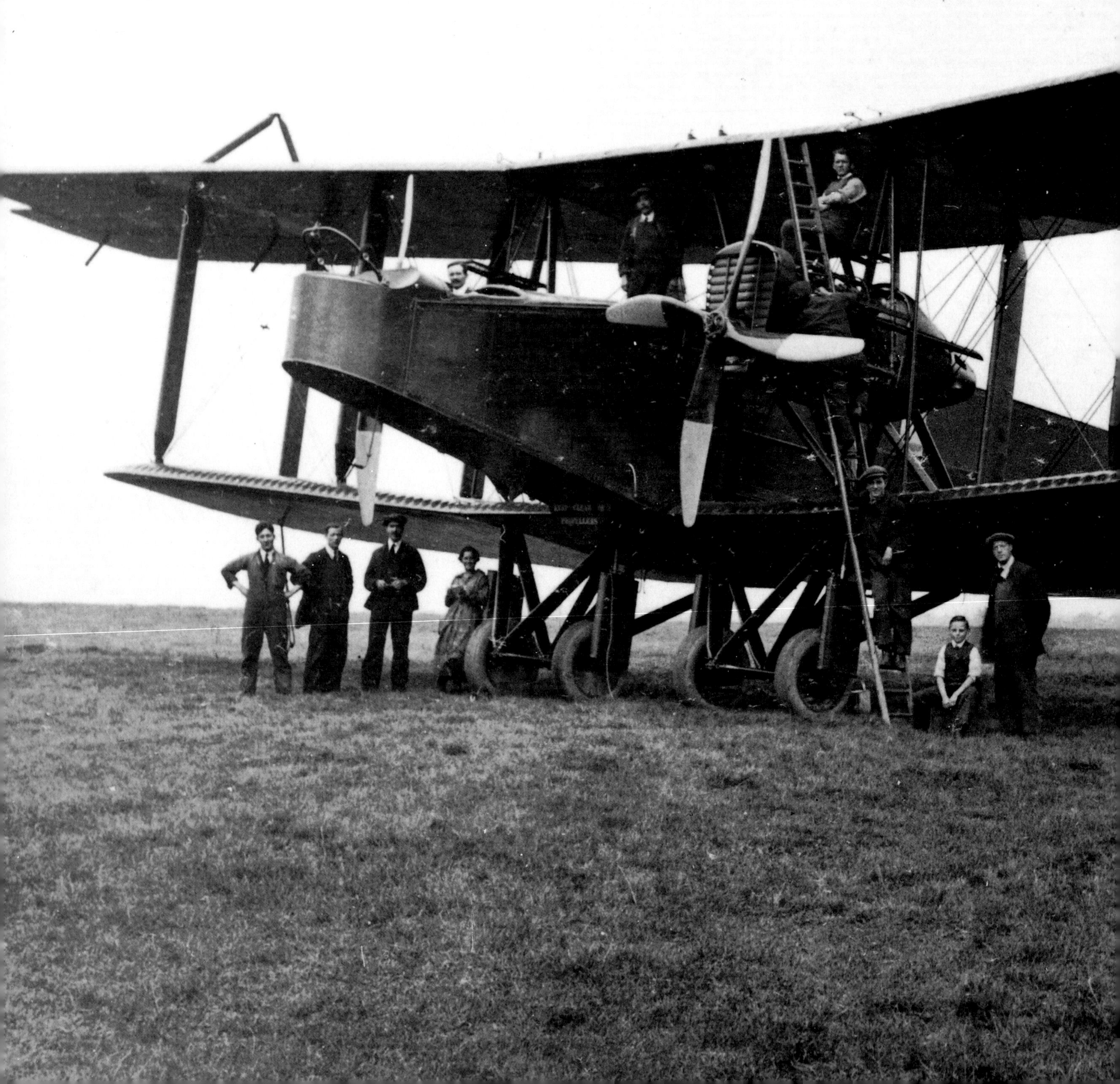

D.8311

One of the most famous fighters of all time, the Spitfire was noted for its curving elliptical wings. From 1940 the outline was altered by adding 20mm cannon, and from 1941 the wingtips were often removed (as here) for better low-level agility.

French air force by one of the most famous pre-war pilots, Roland Garros, and fitted to his previously unarmed Morane-Saulnier monoplane. With a machine gun firing directly ahead Garros shot down five German aircraft in his first three weeks. Then, unfortunately, he was shot down by fire from the ground on April 18, 1915.

The Germans examined his wedges with interest. They were handed to aircraft designer Anthony Fokker with the request that he should produce a copy. His engineers did better than this: they perfected an interrupter mechanism very like some that had been submitted to the arms procurement people of both sides three years earlier and had been ignored. Fokker tested it extensively: he was even required to put on uniform and try it out against Allied aircraft. It was then put into mass production. The first aircraft to be fitted with it was the Fokker E series, the 'E' standing for *Eindecker* ('monoplane'). Though otherwise a quite ordinary aircraft, with an 80 horsepower rotary engine (a German version of the famous French Gnome) giving a maximum speed of about 85mph, when fitted with a fixed machine gun firing ahead through the propeller the E.I became the most deadly killing machine in the sky. Its first victory was scored on August 1, 1915. Very quickly, individual German pilots began to stand out from the crowd, most notably Oswald Boelcke and Max Immelmann. As their scores mounted they became famous as the first 'aces'. More importantly, they thought through and analyzed the whole process of air fighting and formalized it into easily understood rules which could be learned by others.

Until almost the middle of 1916 'the Fokker scourge' caused terrible casualties to British and French pilots, who were shot down in droves in their relatively defenseless aircraft. Only gradually were good Allied fighters developed; but by this time the relatively feeble Eindecker had itself given way to more powerful machines. One of the most agile British fighters was the Sopwith Triplane, so-called because, in order to limit its wing span and increase its agility, it had three small wings. In fact, there was nothing special in having three wings, but this aircraft caused such a stir in Germany and Austro-Hungary that no fewer than 14 companies tried to produce copies. By far the most famous was the Fokker Dr.I, 'Dr' from *Dreidecker* ('three-wings'). The greatest ace of World War I, Baron Manfred von Richthofen (80 victories), was flying a scarlet Dr.I when at last he was shot down on April 21, 1918.

By then most pilots considered the triplane outmoded and regarded the shapely Albatros D.III and Fokker D.VII as the best German fighters. The Albatros first saw action in January 1917 and by the spring all 37 German fighter squadrons on the Western Front were fully or partly equipped with it. The Albatros played the central role in the shooting down of so many RFC machines that Bloody April became a byword throughout the Allied squadrons. The later D.VII did not get into action until April 1918, but it was so formidable that it was specifically named to be destroyed or surrendered in Article IV of the Armistice Agreement after the war.

On the Allied side large numbers of excellent fighters eventually redressed the balance, notably the SPAD, Nieuport, S.E.5a, Bristol F.2b and M.Ic and, above all, the Sopwith Camel. Short, stumpy and hunchbacked (because of the hump caused by the two Vickers machine guns ahead of the windshield), the Camel was tricky to fly, and in fact killed many inexperienced pilots. But those who mastered it found it almost unbeatable. Until 1980 aviation historians agreed that the official total of 1,294 victories made the Camel the top-scoring Allied fighter. Only then did a British author, researching for a book about the Camel, add up the confirmed victories of the British squadrons and found these alone came to considerably more than 2,800; and hundreds more victories were gained by Camels of Allied air forces.

Throughout the war considerable numbers of large and impressive landplanes were used for bombing, while flying-boats ranged far out over the sea, trying to locate enemy fleets and looking for submarines to sink. The British heavy bombers were made by Handley Page, who had been asked by the Royal Naval Air Service to provide them with a 'paralyser of an aeroplane'. The result was the O/100, which by the end of the war had evolved via the O/400 into the V/1500, designed to carry a heavy bomb load from England or France as far as Berlin. Apart from their size and weight the Handley Pages were fairly conventional aircraft. Their German counterparts, called R-aircraft (from *Riesenflugzeug*, 'giant aircraft'), had to meet the difficult requirement of giving access for mechanics to look after their engines in flight. These owed their existence to the farsighted Count Ferdinand von Zeppelin who, on the outbreak of war, asked one company to make a large landplane bomber and another a huge ocean-patrol flying boat. Some of the R-giants had engines in capacious nacelles, with pusher or tractor propellers (or both), and with cabins in the nacelles for the mechanics who also manned machine guns. Most adopted the more difficult solution of putting all the engines within the fuselage in a hot and deafening 'engine room', with gears and shafts driving the huge propellers out either between or on the wings.

After 1918 the needlessly complicated 'engine room' concept went out of favor. Instead the effort went into making engines that were more powerful and, importantly, more reliable, so that they did not need attention in flight. By 1930 Italy had produced the Caproni Ca 90 heavy bomber powered by six engines of 1,000 horsepower each, enabling it to fly with 15,500lb of fuel and up to 17,637lb of bombs. Either of these figures was about the same as a fully loaded Sikorsky or Handley Page heavy bomber of 1918. But sheer size was not necessarily the best answer. A far better bomber than the huge Caproni was the Soviet ANT-6, or TB-3, first flown in 1930.

This bomber, designed and built by a group led by Andrei Tupolev, pioneered the modern configuration in having four engines mounted on the front of an unbraced cantilever monoplane wing. This in turn was made possible by the fact that the TB-3 was skinned not in fabric but in corrugated aluminum. The corrugations were arranged fore-and-aft, and it was not until much later that it was realized that the air flows in many unexpected directions over aircraft surfaces and that corrugated skin causes considerable extra drag. Apart from that the TB-3 was a great machine, 818 being produced in successively improved versions by 1937.

The Stressed-Skin Revolution

The TB-3 was one of the best representatives of an intermediate technology in which the fabric-covered biplane, festooned with struts and wires, was replaced by a clean monoplane exterior which did not, however, make the most of an advance in flight performance. The most important step forward at this time was stressed-skin construction, in which the main stresses are borne by the smooth metal skin covering the wings, fuselage and tailplane. This form of construction had been pioneered by Germans, most notably Dr Adolf Rohrbach, in World War I.

In the 1930s most aircraft manufacturers just went on making fabric-covered biplanes, but in the United States Rohrbach's ideas had not been ignored. Builders such as Northrop, Boeing and Martin recognized that, if they used stressed skin, a cantilever wing could be much thinner than before. As a direct consequence drag could be reduced so drastically that, for any given engine power, airplanes could fly from 50 to 100 per cent faster. In turn, this meant that it would be worth bringing in many new ideas previously rejected as not worth the complication. These included properly cowled engines and radiators, variable-pitch propellers, enclosed cockpits, retractable landing gear, internal bomb bays enclosed by doors, and hinged flaps on the wings that enabled the aircraft, in spite of its increased weight, to land at traditional speeds and pull up on the small grass fields.

The first of the new stressed-skin bombers was the Boeing B-9, which flew on April 13, 1931. This was quite a small machine, but with a top speed of 188mph it could out-perform every previous bomber. But Martin soon produced an even better bomber, which went into production as the B-10 and caused a great stir around the world. In Britain, for example, the newest fighter, the fabric-covered biplane Bristol Bulldog, had a maximum speed of about 168mph, whereas the production version of the B-10 could reach 213mph. One of the B-10's most radical features was the transparent rotating gun turret in the nose. Not surprisingly, the Martin appeared in many versions, and customers were quick to place orders as soon as the U.S. government released it for foreign sales. Yet so fast was the pace of technology that by World War II this pioneer modern bomber was completely obsolete.

One reason for this was the rapid development of fighters. In the mid-1930s many new types of fighter appeared with stressed-skin construction, which gave them better aerodynamic properties even than the special racing aircraft of the 1930s. Among the most important of these mid-1930s fighters was the Curtiss Hawk series of monoplanes, the Supermarine Spitfire and the Messerschmitt Bf 109. All were to be made in enormous numbers and progressively developed throughout World War II. In 1939 they were joined by the Japanese Mitsubishi A6M Zero. This fighter was in combat over China in mid-1940, and some were captured and tested by American, British and other personnel fighting on the Chinese side. Amazingly, neither Washington nor London showed any interest, and when the Zero was encountered at Pearl Harbor on December 7, 1941, it came as a terrible shock to find a Japanese fighter superior to any Allied aircraft then available in the Pacific theater.

One particular attribute of the Zero was its tremendous range. While Spitfires, for example, might have a range of 300 miles, the A6M could, with confidence, be flown 1,000 miles, so it kept on appearing in theaters where no Japanese fighters had been expected. Only gradually was it realized that range is very important, even in a small fighter. When in 1943 the U.S. 8th Air Force really got into its stride, making daylight raids on Germany from bases in England with B-17 Fortresses and B-24 Liberators, it was obvious that, even though these bombers carried powerful defensive armament, their casualties would be greatly reduced if they could be escorted by fighters. No British fighter had sufficient range, but the USAAF Lockheed P-38 Lightning and Republic P-47 Thunderbolt, both exceptionally large, heavy and powerful machines, could accompany the bombers on many missions.

Back in November 1940 North American Aviation, which had never before built a fighter, flew the prototype of a completely new fighter called the Mustang. It had been designed for the British, and at first nobody realized what an exceptional aircraft it was. It was the same size as the Spitfire and other fighters, and had similar power. The startling features were, first, that at low altitude it was faster than a Spitfire (or almost anything else), and, second, that it had three times the fuel capacity and thus three times the range. Soon it was fitted with the British Rolls-Royce Merlin engine, as used in the Hurricane, Spitfire, Mosquito and Lancaster, and from then on the Mustang was one of the best fighters in the sky. With even greater fuel tankage, it accompanied the big U.S. bombers all the way to the most distant targets, such as Berlin and Prague. Marshal Hermann Goering, leader of the German Luftwaffe, later said: 'When I saw those Mustangs over Berlin I knew the war was lost.'

It was near the end of the war, when the Luftwaffe had been almost driven from the skies, before the British heavy bombers – the Lancaster, Halifax and Stirling – operated during the day. In a bitter war in which more than 55,000 British bomber crews were killed, enormous tonnages of bombs were rained on targets all over Nazi-held Europe. It was a campaign marked by the use of a formidable array of electronic gadgetry to help the RAF navigate at night, find targets and confuse the enemy and, from the German side, to confound the British

Continued on page 47

Since 1970 much effort has been expended in restoring a handful of the 10,449 'Zero' fighters of the Japanese Navy. This pristine example is an A6M5, the last of the major production versions.

Two of the most famous American warplanes of World War II: the Boeing B-17 bomber and the North American P-51 Mustang fighter. Both were built in vast numbers (over 12,700 B-17s and over 15,500 P-51s) and served all over the world.

A
232076
E
LL E
LH R

efforts and track down the bombers. The electronic war began slowly in 1942, expanded enormously in 1943, and by 1944 had come to dominate the whole subject of air warfare, especially at night. Ever since, the electronic performance of a combat airplane has been at least as important as its speed or maneuverability. As we shall see, in recent years the subject has been widened to include what is popularly called 'stealth'.

The Coming of the Turbojet

Apart, perhaps, from stressed-skin all-metal construction, the most important single advance in aeronautics since the Wright brothers has been the invention of the jet engine. Of course, strictly speaking rockets are jet engines, but they have found only extremely limited application to aircraft propulsion. This is mainly because the rocket has to store on board all the material needed to form its propulsive jet. This enables it to work in outer space – indeed rockets work much better in space than in the atmosphere – but the penalty is that rockets soon run out of propellants and so are useless for any aircraft that needs to fly for long distances through the atmosphere. The only kinds of aircraft with rocket propulsion that have made sense so far have been ultra-fast research aircraft and target defense interceptors.

The latter are fighter aircraft whose task is the defense of a particular local area, such as a town or factory. Their mission is to take off as enemy bombers approach, climb swiftly and steeply to their level, shoot one or more down and then immediately return to base. Sometimes the whole mission would be over in five minutes. Such aircraft attracted a lot of attention in France and Britain in the 10 years immediately following World War II, not because such aircraft made military sense but because Germany had developed such machines during the war. The most important of these was the Messerschmitt Me 163B Komet (Comet). A quite small, tailless aircraft, this had a swept-back wing mounted in the mid-position and covering almost the whole length of the stumpy fuselage. Most of the fuselage was occupied by tankage holding some 340 gallons of rocket propellants (as a comparison, a Spitfire I held 85 gallons of gasoline). The problem with the Komet was that its two liquid propellants were violently reactive. When brought together in the rocket thrust chamber they gave tremendous power; but if by mischance they mixed anywhere else – even a few drops might be enough – the result was catastrophe.

Although the Me 163B proved almost impossible for Allied fighters to intercept, it killed far more of its own pilots than those of the enemy. Another of its problems was that it took off from a trolley, which was jettisoned soon after take off: sometimes the trolley bounced and hit the Komet. The aircraft could climb to 30,000 feet in not much more than two minutes and on the level could reach 550mph. In the wing roots were two powerful 30mm cannon, one good hit often being enough to destroy an enemy. But then the dangerous part began. Landing was fraught with difficulties because the pilot had to get it absolutely right first time, landing on the correct spot dead into the wind and staying upright on a narrow skid built into the fuselage. Often the jolting of the skid over the rough airfield grass would cause the last remnants of the propellant liquids to mix.

An even more radical target-defense interceptor was the Bachem Ba 349 Natter (Viper). This was almost a piloted missile, because it was fired almost vertically from a ramp. Climbing at 37,000ft/min, it would be leveled off by its pilot who would then have not minutes but seconds to find the enemy bombers, aim the aircraft and fire a sàlvo of rockets carried in the nose. This done, the pilot would pull a lever which would eject him for descent by parachute. At the same time the rear fuselage and tail would separate from the rest of the aircraft, so that the rocket engine (the same as that used in the Me 163B) could be recovered and fitted to a fresh nose and wing. The Ba 349 was a typical 'last-ditch' idea – a desperate attempt to stop the armadas of Allied bombers.

Such rocket-aircraft have found no place in present-day aerial warfare, in contrast to conventional jet aircraft. The turbojet was invented by a young RAF officer, Frank Whittle, when he was serving at Wittering in 1929. It was a very simple idea yet no-one had thought of it before. Many inventors had devised various forms of gas turbine, in which air is drawn in and compressed by a compressor, mixed with fuel and burned to produce a powerful flow of hot gas to drive a turbine. In previous gas turbines the power from the turbine had been used not only to drive the compressor but also to rotate a shaft (as in a factory), to power a locomotive or to turn a propeller. Whittle's brilliant idea was simply to use the three basic elements: compressor, combustion chamber and turbine. The whole residual energy in the flow of hot gas was to be allowed to escape through a nozzle. Whittle saw that such an engine could drive an airplane at any speed, even faster than sound (provided the aircraft drag was low enough and the engine thrust great enough), whereas propellers limited aircraft speed to about 500mph and usually considerably less.

Whittle was an amazingly capable engineer as well as a brilliant test pilot. By 1930, when he took out the patent for the idea of a turbojet, he had worked out the actual design of an engine and completed all the aerodynamic and thermodynamic calculations. Many years later another great engineer, Sir Stanley Hooker, said 'I was awed at the efficiency of Whittle's compressor. I made it worse on one occasion, but I was never able to make it better.' Whittle's unique vision led him to several further inventions, which he patented in the mid-1930s.

Continued on page 50

***Top left*: A simplified cut-away showing how most of the stumpy fuselage of the Me 163B was filled by the rocket engine and its propellant tanks.**
***Left*: Removing the BMW 003B turbojet from an He 162 fighter in a hangar at Farnborough, England, in 1945.**

13

A Messerschmitt Me 262A-la twin-jet fighter about to taxi out on a mission towards the end of World War II. Apart from the slower Meteor I, the Allies had nothing that could meet the 262 on anything like even terms.

Despite its enormous cost, and the fact that in post-war years defense funds were very limited, the USAF bought 385 of these gigantic Convair B-36 heavy bombers. This version, the B-36D, added four jets under the outer wings.

One was the afterburner, a special form of enlarged jetpipe in which additional fuel can be burned downstream of the engine to give extra thrust, especially for supersonic speed. Another was the turbofan, in which extra air compressed by the compressor is bypassed around the rest of the engine to give a relatively cool and slow propulsive jet surrounding the fast and very hot jet from the core (the basic turbojet). The turbofan dramatically reduces noise and can also reduce fuel consumption by about half, but it is best suited to subsonic aircraft such as airliners.

However, no-one in the British Air Ministry or RAF was in the least bit interested in such advanced ideas as turbofans and afterburners: no-one even wanted to know about the basic idea of the turbojet, and Whittle had to let his patent lapse. Six years after his invention a German, Dr Pabst von Ohain, had the same idea. He was luckier: aircraft designer Ernst Heinkel, mad on speed, immediately gave von Ohain a job and said: 'Make me a jet aircraft'. The result was the He 178, first flown just before the war on August 27, 1939. Heinkel would also fly the world's first jet fighter, the He 280, on April 5, 1941; but this never went into service.

The first jet fighter to join a combat unit was the Gloster Meteor I, which entered service with RAF No 616 Sqn in July 1944. The German Messerschmitt Me 262A-1a followed about two months later. Little effort, however, was made to speed Meteor deliveries to the RAF and only a little over 50 had become available by the end of the war. In contrast, more than 1,400 of the Me 262s were built. In some respects the Me 262 was an outstanding aircraft, with a speed of 540mph and a devastating armament of four 30mm cannon. But it lacked maneuverability except at low speeds, accelerated slowly, had unreliable engines and other problems, and many of them crashed.

Back in 1940, engineers at the German Arado works had perceived that jet propulsion would make a long-range reconnaissance aircraft virtually immune from interception. The result was the Ar 234 Blitz (Lightning). The first prototype flew on June 15, 1943. The original idea of taking off from a jettisonable trolley and landing on a skid was abandoned, and the Ar 234B production versions had normal landing gear. These aircraft not only transformed the Luftwaffe's ability to take reconnaissance photographs together with some ME 262s; they were also the first jet bombers, seeing much action from the autumn of 1944 and carrying bomb loads of up to 2,205lb.

106

***Above*: Two Mig-15s of the Czech Air Force.**
***Right*: This F-86 Sabre was made in Canada for the RAF.**

Perhaps the most remarkable military-aircraft development program of all time concerned the Heinkel 162A, popularly called the *Volksjäger* (People's Fighter). This was the ultimate 'last-ditch' concept and codenamed the Salamander program: its aim was to restore to the Luftwaffe its lost command of the skies. In the last 90 days of 1944 the program was drawn up and approved, the He 162 designed, built and flown, and plans made for mass-production at the rate of 4,000 per month. The plan was to put swarms of *Volksjägers* into the sky in the spring of 1945, flown by hastily trained Hitler Youth pilots. The scheme was unrealistic because the He 162 needed a skilled and experienced pilot. But the aircraft itself was a remarkable achievement. Its small but extremely efficient airframe was made largely of wood. A single turbojet was mounted on top of the aircraft, the jet passing between the twin fins. As the jet inlet was immediately behind and above the cockpit, the pilot was provided with one of the first ejector seats, which was fired out of the aircraft by an explosive cartridge. Armament consisted of two 20mm or 30mm guns, and maximum speed was 522mph or – using a special high thrust rating permitted for 30 seconds – 562mph. But the war ended before the *Volksjägers* could have any effect.

Supersonic Flight

In the years immediately after the war most research was directed into putting turbojet technology to use for both military and civil aviation; and soon aeronautical engineers took an active interest in the problems and possibilities of supersonic flight.

Britain lagged behind the United States in the development of military aircraft, so that when the Korean War broke out in 1950 the only Allied fighter able to compete with the opposition in Korea was the superb North American F-86 Sabre, designed by the team that created the Mustang. It was the first swept-wing jet fighter, and it pioneered many other developments, even though it first flew as early as October 1, 1947. From the start it was an absolute winner. In the spring of 1948 its pilots found that, by rolling inverted and then pulling through the second half of a loop at full throttle, they could exceed the speed of sound. No unpleasant handling problems resulted, and what the media called 'piercing the sound barrier' was soon routine for hundreds of regular squadron pilots, to the accompaniment of the characteristic sonic booms heard on the ground. The armament of early versions of the Sabre was six 0.5in machine guns, exactly the same as in most Mustangs. In March 1951, however, deliveries began of the F-86D, a totally redesigned all-weather and night fighter characterized by a search radar in the nose, displacing the engine inlet to the 'chin' position. To enhance the flight performance despite a great increase in weight, the engine had an afterburner; and instead of guns the F-86D carried a battery of rockets. It was the first single-seat radar-equipped interceptor, and 2,504 were delivered of this version alone. Altogether, including carrier-based Fury versions for the U.S. Navy, some 9,000 Sabres were built, more than for any post-1945 fighter outside the Soviet Union.

It was just as well that the Americans had something as good as the F-86, because their pilots were soon encountering swept-wing fighters bearing the red stars of North Korea or China. It was soon apparent that these agile machines, hitherto unknown in the West, were more than a match for every Allied fighter except the Sabre. Even the latter was inferior in rate and angle of climb, in performance at extreme altitude and in several other respects, and it notably lacked the punch and long range of the opposition's heavy-caliber cannon. The unwelcome adversary was the Soviet MiG-15, made possible by the British Government's inexplicable gift to Moscow of the very latest and most powerful British engine, the Rolls-Royce Nene. It also powered many other Soviet aircraft, including the Ilyushin 28 jet bomber: and an improved version powered the later MiG-17 fighter.

XB982

N
XF523

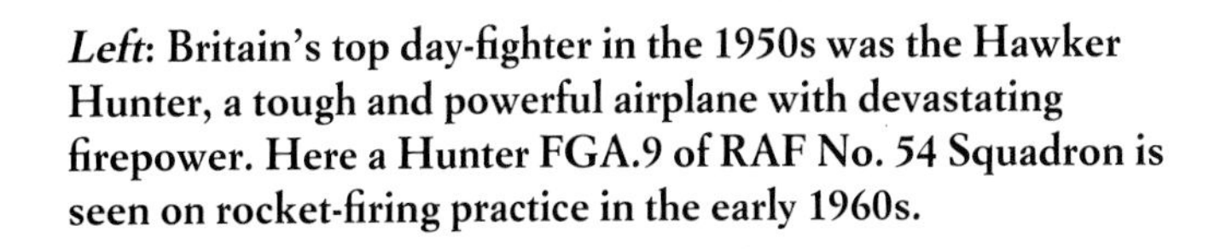

***Left*: Britain's top day-fighter in the 1950s was the Hawker Hunter, a tough and powerful airplane with devastating firepower. Here a Hunter FGA.9 of RAF No. 54 Squadron is seen on rocket-firing practice in the early 1960s.**

Until the Korean War Soviet warplanes had been either ignored or underrated in Britain, but the MiG-15 administered the same kind of salutary shock as the Japanese Zero back in 1941. Suddenly it was recognized that the Western Allies, who in 1949 had formed the NATO alliance, enjoyed no monopoly of first rate military aircraft. Indeed, in the 1950s the pendulum if anything swung the other way, even in the matter of strategic bombers.

The Big Bombers

By far the most advanced strategic bomber of World War II was the Boeing B-29 Superfortress. Planned in 1939, when its ancestor the B-17 was just getting settled down in service, the B-29 first flew on September 21, 1942. It added roughly 50 per cent to the speed and over-target altitude of heavy bombers, and its pressurized cabins brought previously unknown comfort to the crew. A major innovation was that the heavy defensive armament of cannon and 0.5in machine guns was mounted in powered turrets controlled by gunners in sighting stations elsewhere in the aircraft. The unprecedented range of the B-29 enabled heavy attacks to be mounted against the previously too-distant Japan: the two nuclear weapons dropped by B-29s on August 6 and 9, 1945, on Hiroshima and Nagasaki respectively, abruptly ended the war in the Far East.

The Soviet leader Josef Stalin did everything he could to get hold of some B-29s, without success. He need not have bothered because in 1944 three made forced landings in Siberia. Immediately they were taken apart and minutely examined, their materials analyzed – and a Soviet copy was rushed into production as the Tu-4. This was a staggering achievement, bearing in mind the incredible complexity of such aircraft. In turn the Tu-4 was developed via several other aircraft into a gigantic machine with swept wings and four turboprop engines.

First flown in 1954, this huge aircraft puzzled Western analysts, who argued that there was no point in having swept wings and tail in an airplane driven by propellers. They overlooked two things: the NK-12 engine was by far the most powerful engine ever developed to drive propellers, starting life at 12,000 horsepower and soon delivering about 15,000; and the extraordinary eight-blade contra-rotating propellers of

Continued on page 58

In the Western world two of the pioneer Mach 2 fighters: the Lockheed F-104 Starfighter and McDonnell F-4 Phantom II, both powered by the General Electric J79 engine (the F-4 with two J79s). Here an F-104C escorts a camouflaged F-4C.

these aircraft were set to such a coarse pitch that even at 500mph they were rotating quite slowly. Designated Tupolev Tu-20, Tu-95 and Tu-142, depending on the version, the aircraft was dubbed 'Bear' by NATO. For years it was grossly undervalued in Washington, while frantic measures were taken to counter the supposed threat of the 'Bear's' partner, the Mya-4 'Bison'. Smaller than the Tupolev, the Mya-4 was powered by four enormous turbojets buried in the roots of the swept wings. Such was the American fear of the 'Bison' that the fighter and missile defenses of the United States were rapidly augmented, at enormous cost, while the USAF Strategic Air Command (SAC) was swiftly re-equipped with bombers and air-refueling tankers to provide almost global airpower.

Strategic Air Command had begun just after the war with the B-29. In August 1947 it began to take delivery of the biggest bomber ever built. The Convair B-36 had a wing span of 230 feet, and not the least of its features was that crewmembers could travel between the front and rear pressure cabins by a railed trolley travelling along an 80-foot tunnel above the vast bomb bays and through the wing. Six 3,500 horsepower Wasp Major piston engines were installed in the wing, driving pusher propellers. From July 1949 additional power was provided in the form of four J47 turbojets in pods hung below the outer wings, to boost over-target height, and also to raise maximum speed from 381 to 439mph. A tiny jet fighter, the McDonnell XF-85 Goblin, was developed to fold up and be carried inside a B-36. Unfolded and released, it was to fight off hostile fighters before hooking back on again and folding away in the giant's bomb bay. One squadron, equipped with the GRB-36D, carried modified Republic RF-84F reconnaissance aircraft to targets thousands of miles beyond their normal radius of action.

Although an intriguing idea, all the 'parasite' concepts involving smaller aircraft hooking back on a large parent have been fraught with difficulty and danger, and they have had quite a short life. On the other hand, air-to-air refueling has grown to play a vital role in the work of several air forces. In the 1920s the pioneers of in-flight refueling simply poured petrol (gasoline) down a hose from one airplane to another flying beneath it. By the 1930s large tanker aircraft were being used to refuel flying boats which could not otherwise have made non-stop journeys across the Atlantic. The technique was cumbersome, however, requiring a long cable and grappling hook to catch a trailed hose and winch it round in a curve until it plugged into a socket on the receiver aircraft.

In the 1950s the method was greatly improved. The hose was reeled out from a drum on the tanker, pulled by the drag of a basket-like drogue into which the receiver simply inserted a forward-pointing probe; the coupling was automatic. This method is today used all over the world. The U.S. Air Force uses a different method in which an operator aboard the tanker 'flies' a rigid but telescopic and pivoted tube, called a boom, until he can 'fire' the end coupling into a receptacle on top of the receiver. This method puts the onus of making a good contact on to the boom operator, and also has the drawback of refueling only one aircraft at a time. A big tanker can have three hose reels at work simultaneously.

Some of the first aircraft to be equipped for flight refueling were the U.S. fighters and bombers of the early 1950s. SAC was supplied with 2,000 Boeing B-47 Stratojets, which looked breathtakingly futuristic when the prototype flew on December 17, 1947. Features included sharply swept wings and tail, six turbojets mounted outside the airframe in underwing pods (then a novel idea, though first used on the single wartime Ju 287), tandem 'bicycle' landing gears and a beautifully streamlined fuselage housing the previously unheard-of fuel capacity of 17,000 gallons. There were no unsightly turrets, just twin 20mm guns in the tail managed by a remote gunner with the aid of radar. Whereas wartime bombers typically had a crew of ten, the B-47 had just three, seated beneath a long 'fighter type' canopy in the nose. It was one of the most beautiful aircraft ever built, but it was not easy to fly. One reason was that, whereas the wing loading (the weight supported by a unit area of wing) in bombers of the 1930s was typically 15lb/sq ft and the very heavy B-29 pushed the limit up to 76lb/sq ft, the B-47 reached the frightening level of 145lb/sq ft. This made the bomber tricky to handle at high altitude, where maximum speed and stalling speed tend to coincide, and after landing, when a huge braking parachute was needed to help the overworked wheel brakes.

After the B-47 came the Boeing B-52. While, even with flight refueling, the B-47 was unable to reach all the major SAC targets, the B-52 was a truly global bomber with all the range needed. The first studies for the B-52 all specified turboprops because it was thought that no jet bomber could possibly meet the requirement of a range of over 6,000 miles. Then in 1950 Pratt & Whitney developed an engine which led to the J57, a very advanced turbojet with separate low-pressure and high-pressure compressors driven by separate turbines through concentric shafts. The high compression this made possible resulted in an engine of unprecedented efficiency, and Boeing redesigned the B-52 as a jet. Nevertheless, although the J57's thrust of 10,000lb was among the most powerful in the world at that time, the B-52 required no fewer than eight engines. They were housed in four twin pods. The aircraft first flew in April 1952. Eventually 744 were built in successively improved versions. The last model, the B-52H of 1961, was powered by eight turbofan engines of 18,000lb thrust. These engines dramatically reduced the noise and increased range considerably. The early versions could carry 37,550 gallons of fuel and could fly 7,370 miles. The B-52G introduced integral-tank wings in which, instead of housing separate fuel cells, the wing itself is sealed to serve as a huge tank. It carried 47,975 gallons and gave a range of 8,406 miles. The B-52H, however, could fly 10,130 miles on the same amount of fuel.

The English Electric Lightning F.6., seen here firing a Red Top missile, was withdrawn in 1988.

Following the original Sea Harrier, which played a vital role in the Falklands campaign in 1982, British Aerospace is now delivering the Sea Harrier FRS.2. This view shows the new nose radar and advanced medium-range air-to-air missiles.

The B-52 is remarkable in many ways, not least for its longevity. Today many B-52G and H bombers, some modified to carry ALCMs (air-launched cruise missiles), supersonic SRAMs (short-range attack missiles), or Harpoon anti-ship cruise missiles or ACMs (advanced cruise missiles), seem destined to continue in service until at least 50 years after the prototype first flew! In contrast, the Convair B-58 Hustler, the world's first supersonic bomber, had quite a short career. First flown on November 11, 1956, the B-58 was a bold response to a formidably challenging requirement. Whereas the makers of the first jet bombers had found it difficult to get sufficient range, the idea of a supersonic bomber seemed ridiculous because, owing to the much greater drag, its range would be roughly half its subsonic equivalent. Convair cheated by designing the XB-58 to be quite a small aircraft, with a span of only 56ft 10in. It was able to be so small because most of the disposable fuel and bomb loads were carried externally in a gigantic pod slung under the slim fuselage. Various pods were produced, some carrying the bomb recessed underneath, some having the bomb in the nose of a fuel pod and some even having their own rocket engine. After dropping the pod the B-58 was almost halved in size, though at take-off it weighed 163,000lb and could fly 5,125 miles without air refueling. The B-58 entered SAC service in 1960, but was withdrawn just 10 years later.

Fighters: the New Generation

Supersonic fighters presented a much easier problem. The USAF called their early examples the Century Series because the first was the North American F-100 Super Sabre, followed by the McDonnell F-101 Voodoo, Convair F-102 Delta Dagger and Lockheed F-104 Starfighter. All had long careers; indeed, the F-104, one of the most easily recognizable aircraft because of its tiny sharp-edged wing, is still the standard all-weather interceptor of the Italian air force.

The Soviet Union produced an even wider variety of supersonic fighters. The MiG-21, which first flew in 1955, has been made in possibly greater numbers than any other jet fighter, and developed versions are still in full production in China. The MiG-21 is a tailed delta; in other words, it has a triangular wing and also a horizontal tail. Another very

A joint product of McDonnell Douglas, St Louis, and British Aerospace, the Harrier II is a major improvement over the original version. One advantage is that it can either carry twice the warload or fly twice as far. This example is a Harrier GR.5 of the RAF. Other users are the U.S. Marines and the Spanish Navy.

B
ZD324

Below: The Mirage 2000C and (*right*) the Mirage F1. C-200, both by Dassault-Breguet.

The Grumman swing-wing F-14 Tomcat first flew in 1970 yet it still has qualities which even today cannot be matched by other fighters.

NAVY
VF-32

When two Soviet MiG-29s – a 29A fighter and 29UB trainer – visited England in 1988 they were escorted by Tornado F.3 interceptors of the RAF. While the F.3 is a long-range two-seater the MiG-29A is a brilliant close-combat fighter.

successful family of fighters, likewise first flown in 1955, is the French Dassault Mirages. These have no horizontal tail, the wings carrying dual-purpose control surfaces called elevons. The latest version, the greatly updated Mirage 2000, is still in production, Dassault has also produced a completely different family called the Mirage F1 series, with a high swept wing and normal horizontal tail. Perhaps the most striking of these early supersonic fighters was the Swedish Saab 35 Draken, with a so-called 'double delta' layout in which most of the fuel, armament and equipment was housed in what could be considered as either the outer parts of the fuselage or the inner parts of the wing.

On May 27, 1958 McDonnell Aircraft flew the prototype F-4 Phantom II. Not asked for by any customer but designed in collaboration with the U.S. Navy, the Phantom was hardly beautiful, but its performance was exceptional. It was powered by two General Electric J79 engines (which also powered the F-104 and B-58). Like the F-104, the big Phantom had blown flaps: hot compressed air from the engines, expelled at supersonic speed across the lowered flaps, prevented 'break-away' of the airflow and greatly reduced landing speed. Despite being burdened by all the extra equipment and structural strength needed for operation from aircraft carriers, the Phantom set world records for speed (over 1,600mph), rate of climb and various other categories, besides having up to eight air-to-air missiles and the best and most powerful radar of any fighter of the 1950s. Almost unprecedentedly, the U.S. Air Force decided to adopt this Navy aircraft, and altogether 5,211 were built.

In 1964 the British government decided to buy Phantoms for the Royal Navy instead of the naval version of the Hawker P.1154. A year later the RAF version of the P.1154 was also canceled in favor of Phantoms. The decision appeared curious, not only because it seemed to threaten the future of the British aircraft industry but also because it meant replacing a STOVL 'short take-off, vertical landing' aircraft by a CTOL 'conventional take-off and landing.' No CTOL combat aeroplane can operate unless it has a long paved runway or a huge aircraft carrier equipped with catapults and arrester gear. What was especially strange about buying Phantoms for the Royal Navy was that the British government canceled plans for a new aircraft carrier and proceeded to eliminate the Royal Navy's carrier force.

53
10

After watching this Soviet Sukhoi Su-27 fighter perform at the 1989 Paris airshow many Western observers have agreed that the West has nothing that could "stay with it" in a dogfight.

The advantage of the STOVL airplane is that it can operate from any small, flat space, such as on top of containers on a merchant ship, on the small deck at the stern of a warship, in a forest clearing or a yard behind a supermarket. In any future war any military target of any importance whose position is fixed and known to the enemy is almost certain to be destroyed by missiles within the first few minutes of hostilities being declared. Obvious examples of such a target are airfields. STOVL aircraft do not need airfields. Some 30 years ago this unique advantage was considered to be of great significance; today it is ignored by almost all air forces. This is partly because the only combat-ready STOVL aircraft are British or Russian. The Soviet types are the Yak-38 and Yak-41 of the Soviet Navy, and the British ones are the various forms of the British Aerospace Harrier. The latter are especially attractive in having only one engine. This engine, the Rolls-Royce Pegasus, was the brainchild of engineers at Bristol in the late 1950s, who realized the advantages of using a single powerful turbofan fitted with four swivelling nozzles. The left and right front nozzles discharge fan air and the left and right rear nozzles discharge the hot core jet. All four are mechanically linked so that they point downwards for a vertical take-off, hovering or landing. To accelerate forwards the nozzles are gradually rotated to point to the rear, lift being transferred from the four jets to the wings. The nozzles can also be vectored in flight to perform 'impossible' maneuvers, making the Harrier among the most agile and difficult adversaries in air combat. The original Harrier entered service with the RAF at the beginning of 1969, mainly in the ground attack and reconnaissance roles. The better-equipped Sea Harrier reached the Royal Navy in 1979, and without it the recapture of the Falklands Islands in 1982 would have been impossible. Today, thanks largely to McDonnell Douglas in St Louis, a new-generation Harrier family is in production with greater range and weapon load, vastly better equipment for night and all-weather operation, and a totally new standard of safety and survivability.

Back in 1960 an oil-and-water mixture of politicians and visionaries tried to give the U.S. Air Force the ultimate in new fighters. Originally called the TFX (Tactical Fighter Experimental), it demanded an 'all-singing, all-dancing' aircraft with, at that time, barely credible capabilities. The most severe requirements were that it must operate from a 3,000-foot unpaved airstrip and attack a target 800 nautical miles (920 miles) away with a 10,000lb bombload after flying the whole distance at treetop height; the final 200 nautical miles (230 miles) in each direction being flown at Mach 1.2. Another requirement was a ferry range of 3,000 nautical miles without air refueling. One of the politicans – Secretary of Defense Robert S. McNamara – decided much money would be saved if the USAF and U.S. Navy bought versions of the same aircraft. The result, the F-111, was not at all what had been ordered. The aircraft had afterburning turbofan engines, a terrain-following radar to allow it to follow the undulations of the ground automatically to avoid detection by hostile radar, and the new development of pivoted 'swing wings', spread out to a wide span for low speeds and progressively folded back for supersonic flight. The F-111 was the best tactical bomber in the world – which was fine, except that the U.S. government had asked for a fighter. Far from having a gross weight of 'approximately 60,000lb' as specified, the F-111 in its various versions weighed from 92,500 to 114,300lb. Early in the program the Navy pulled out. Instead of well over 2,000 (not including the predicted exports), only 562 F-111s were produced.

The U.S. Navy plumped for the Grumman F-14 Tomcat, which likewise had two crew, two similar Pratt & Whitney engines, and swing wings. It, too, was on the heavy side, with a maximum weight of 74,349lb; but its combination of range, performance, radar and avionic capability and, especially, three different kinds of air-to-air missiles as well as an internal gun, have kept it in the front line for nearly 20 years. The USAF counterpart is the McDonnell F-15 Eagle, with one seat, a fixed wing of large area (608sq ft) and much newer and better engines. Today what may be the ultimate Eagle, the F-15E, has two seats, augmented avionics and the ability to deliver heavy bombloads on ground targets (a mission explicitly excluded from the original F-15 specification).

In the 1970s the concept of an LWF (lightweight fighter) emerged, and a competition was won by the General Dynamics F-16, again with a fixed wing and a single engine of the type fitted to the twin-engined F-15. At first the general opinion was that the LWF was cheap and therefore inferior and that it would never be bought for the USAF. In fact the F-16 proved to be such an impressive fighter that it has been bought in much greater numbers than the F-15. A twin-engined rival, the McDonnell Douglas F/A-18 Hornet, was developed for the U.S. Navy and, like the F-16, has been widely exported.

The corresponding Soviet fighters, designed when all details of these American fighters were known, are the Mikoyan MiG-29 and Sukhoi Su-27. Aerodynamically both are similar; like the F-15 they have a large fixed wing, twin engines and twin fins. They differ in many respects, notably in having unprecedented all-round performance and maneuverability and distinctly superior equipment in the matter of radar, infra-red tracker, laser ranger, missile types and internal 30mm gun. The MiG is a multirole aircraft; the Su-27, with truly astonishing maneuverability, is a pure fighter. Britain's only fighter is a version of the swing-wing Tornado, originally designed as an all-weather attack and reconnaissance aircraft. Two fighter Tornado F.3s escorted two MiG-29s on a visit to Britain in 1988. The F.3s could not have expected to win in a dogfight with the Soviet aircraft. On the other hand, the MiGs could not have flown through a blizzard to Iceland, established the identity of an intruding aircraft, and then flown back!

Often the intruders have been late versions of the remarkable Tupolev 'Bear', which has now been in production for 35 years. Some of these newer versions are for reconnaissance, some are ASW (anti-submarine warfare) aircraft, and others carry cruise missiles or extraordinary radio equipment for communicating with submerged submarines. All have global range. The only U.S. aircraft remotely comparable is the B-52, all of which are old and rather tired.

Stealth Technology

In the 1960s it was decided not to go into production with an amazing bomber called the XB-70 Valkyrie, one of the heaviest, fastest, noisiest and most powerful aircraft ever built. There followed 10 years of study of AMSA meaning Advanced Manned Strategic Aircraft, and not, as some critics asserted, 'America's Most Studied Aircraft'. Eventually, the AMSA was developed into an actual bomber, the Rockwell B-1, first flown on December 23, 1974. Then in June 1977 President Carter canceled the whole program, claiming the B-52 could do the job, carrying cruise missiles. This was obviously nonsense and Rockwell continued to develop the aircraft into the B-1B, which was capable of carrying heavy loads of conventional weapons at treetop height, but unable to fulfil the original requirement of Mach 2 at high altitude. In October 1981 President Reagan ordered 100 B-1Bs, and these were delivered by April 30, 1988. The four-seat B-1B has a swing wing, four powerful General Electric afterburning turbofan engines, a special system to smooth out the ride at full throttle at low level (both to avoid structural fatigue and improve the performance of the crew), and the world's biggest avionic systems for navigation, weapon aiming and, above all, detection and countering of hostile anti-aircraft systems.

An almost exact Soviet counterpart, but even bigger, is the Tupolev Tu-160 (NATO name 'Blackjack'). At first glance, except for being painted in very pale instead of very dark colors, this might be a B-1B. It is almost identical in shape, but the span of the wings (spread out at minimum sweep) is almost 183ft in the Tu-160 compared with just over 136ft, and the Soviet aircraft is 30ft longer at 177ft. Gross weights are 477,000lb for the B-1B and an estimated 630,000lb for the Soviet bomber, making it the heaviest warplane in history.

Impressive as these bombers are, they have to rely on on-board avionics systems to warn them of hostile defenses and protect them against being shot down. Such systems are

***Opposite:* The multirole McDonnell Douglas F-15 Eagle, is the top fighter of the U.S. Air Force.**

The F-16C is the current single-seat production version of the General Dynamics Fighting Falcon.
Inset: One of the 26 super-agile F-16N 'Aggressor' Falcons, with uprated engines, used by the U.S. Navy to act the part of enemy fighters.

exceedingly complex, expensive and heavy; the B-1B system, called ALQ-161, requires about a ton of cables alone, and consumes electric power at the rate of 120kW (120,000 watts). It is obviously far better to try to make the basic aircraft harder to detect. This was tried in 1913 when various warplanes were covered in cellophane-type material in an attempt to make them 'transparent'. In 1935 the 'father of radar', Robert Watson Watt, pointed out that future aircraft should be designed to be as invisible as possible to hostile radars, but nothing was done for another 40 years. Since the late 1970s the USAF has led the way in LO (low observables) technology, popularly called 'stealth'.

When the media discuss 'stealth' they almost invariabley concentrate on the problem of trying to defeat enemy radars. Important as that is, it would be of little advantage to an aircraft that could be clearly seen with the eyes, could be heard from miles away, and pumped out infra-red (heat) from its engines. In fact, reducing the visibility on enemy radars is one of the easiest parts. Making an airplane difficult to see against such widely differing backgrounds as blue sky, white cloud, snow, grass, dark mountains or blue, green or gray sea is obviously not a simple matter. Most difficult of all is the problem of heat. The engines of a modern bomber typically develop about 100,000 horsepower, and this produces enormous heat in the jets. If we bear in mind the claims that today's infra-red detectors can locate a lighted cigarette-end at the range of 30 miles, we can get some idea of what the designers of 'stealth' aircraft are up against.

After tests with a Wittman lightplane and a series of specially designed Lockheed XST (Experimental Stealth Technology) aircraft, the USAF put into service an initial quantity of Lockheed reconnaissance/attack aircraft with the strange designation of F-117A (strange because it does not fit a previous scheme, and the aircraft is not a fighter). This single-seater has an external surface made up almost entirely of hundreds of flat panels, each reflecting light and radar energy in a different direction. A tailless delta aircraft with a vee-type pair of fins, it has two General Electric turbofan engines buried in thick inner portions of the wing. Attack weapons are carried internally.

The next generation is the Northrop B-2, which as its designation suggests is planned to follow the B-1B into service with SAC. First flown on July 17, 1989, the B-2 has been designed and built totally by computer, the electronic data-processing power used being far greater than in any previous project. The result is an all-wing bomber using a 'stealth' technique quite unlike that of the F-117A. Curiously, the B-2 has the same configuration as a Northrop all-wing bomber of 45 years previously (the XB-35 and YB-49), and precisely the same span of 172ft. There the similarity ends. The huge B-2 has a crew of only two, with an occasional third seat. Its engines are four large General Electric F118 turbofans, specially designed to give a cool and quiet jet, and buried in the inner wings (as are the engines of the F-117A) with complex inlet and nozzle features to defeat hostile sensors. The B-2 is a remarkable achievement. The question is: can even the United States afford to buy 132 at $600 million each?

Main illustration: The world's first 'stealth' warplane, the F-117A, called The Wobbly Goblin by its USAF pilots. It first flew in June 1981.
Insets: Two artists impressions of possible ATF (Advanced Tactical Fighters).

USAF Strategic Air Command was given a tremendous boost in capability by the production of 100 Rockwell B-1B bombers, although extra work is being done to make the complicated avionics meet all the requirements.

The biggest bomber in the world, the Soviet Tu-160 looks very much like an enlarged B-1B. Called 'Blackjack' by NATO, it is 177 feet long and is estimated to weight about 300 tons.

The Airbus A320 is one of the most technologically advanced commercial jetliners with a choice of two different engines.

3

WINGS FOR PEACE

WINGS FOR PEACE

We saw in Part 2 how, after showing complete indifference born of ignorance in the years leading up to World War I, both the Allies and the Central Powers poured money into the development and production of warplanes during the years of conflict. As soon as the war was over, however, the situation changed abruptly. In Britain the responsible minister, Winston Churchill, announced with chilling finality: 'Civil aviation must fly by itself'. All over the world for the next 20 years, infant airlines struggled to make ends meet. It was not until World War II started in 1939 that the financial floodgates were opened once more – for military aircraft only, of course.

Today the situation is rather different. Despite having had an uphill struggle for most of the way, the world's airlines have become giants. Thanks almost entirely to the work of the engineers and technologists, commercial transport by air has developed out of all recognition. Everyone expected the airliners to grow bigger, faster and perhaps longer-ranged. What few would have dared to predict in, say, 1955, is the extent to which the cost of air travel has fallen. Lowering prices means that you sell more seats: the number of passengers flying across the North Atlantic has grown since 1945 from dozens to many millions annually. Today the world's major airlines – those belonging to the International Air Transport Association – employ over two million people, and even more are employed by the world's civil airports. It has all become very big business indeed. Yet, strangely, this revolution in transport seems to attract little interest. A standard pocket-book, *The Aviation Enthusiast's Reference Book*, has a single diagram showing the names of the component parts of a light aircraft – but it is otherwise devoted entirely to military aviation.

Pioneering the Air Routes

It was not so 200 years ago, when Cayley dreamed of being able to navigate that ocean 'that comes to every man's door'. There have been many visionaries since, such as Dr Albert Plesman, the founder and leader of the Dutch airline KLM, who in the 1920s pointed out that 'The ocean of the air unites all peoples.' Sad that in May 1940 his country should have been devastated by bombers and he himself imprisoned by the invaders. Yet today the sheer scale of air traffic and the ease with which one can fly to the remotest corners of the world, must be reckoned one of mankind's greatest success stories in the post-war era.

Air travel is so simple and commonplace nowadays that we find it hard to picture air travel in the years immediately following World War I. In hindsight it may seem to have been exciting, even romantic: at the time, however, probably its only good feature was that there were so few passengers that each received individual attention. When the British airline, Aircraft Transport and Travel was formed, it was not able to operate scheduled services until August 25, 1919, but on July 15 a flustered gentleman appeared at their Hounslow terminal (a shed) and asked if he could be taken to Paris. He was the glass magnate Major Pilkington, and he had missed the boat train. AT&T charged him $82 (£50) – equivalent to about $1,625 (£1,000) today – and got him there. Then, and for many years afterwards, every airline passenger was obviously someone out of the ordinary. Some flew merely in order to say they had done it. Hardly anyone flew because it was faster, surer or more convenient; in most cases, it was none of those things. The first regular international service by a totally civil airline is generally held to be that opened on March 22, 1919 between Paris and Brussels. The Farman Goliath airliner used was regarded as a real giant: in front was a cabin for four passengers: at the rear was another cabin for eight. All passengers had wicker chairs (not fixed to the floor), and there were curtains at the windows. Between the cabins was the pilot in an open cockpit. The service operated only once per week, the one-way journey took two hours 50 minutes, and the fare was 365 French Francs. The train for the same journey took three hours; the fare was 55 French Francs and passengers were delivered to the center of the city. But Europe has always been difficult for airlines, with its mixture of bad and unpredictable weather and superb surface communications.

Long intercontinental flights, where the speed of airplanes would have shown to greatest advantage, were beyond the capabilities of the early airliners. Even long overland routes involved frequent stops, at which the passengers often had to change to a different aircraft or even to a different mode of travel. Today, when a Boeing 747 can fly from London, England to Australia non-stop, it is quite a labor of love merely to study the exhausting schedule of 1930 for the journey to Karachi, which is much less than half as far as Australia. Passengers would board an Argosy at Croydon Airport and, with luck (weather and engine failure quite often caused problems), fly to Basle, Switzerland. There a sleeping-car

The nine-winged Caproni Ca 60 Noviplano – one of the most fantastic flying machines ever built.

express train would take them to Genoa, where the next day they would board a Calcutta flying boat. This would make the long flight to Alexandria, Egypt, refuelling at Rome, Naples, Corfu, Athens, Suda Bay and Tobruk (Libya). From Alexandria, passengers would transfer to the land aerodrome and fly in a D.H.66 Hercules to Karachi, refuelling at Gaza, Rutbah Wells (Iraq), Baghdad, Basra, the Iranian Gulf, ports of Bushire and Lingeh, Jask and Gwadar (near the present frontier between Pakistan and Iran). Total journey time to Karachi: seven days. This is a true measure of what could be done in 1930.

The Early Visionaries

Aviation has always had its full share of dreamers, and some of them have let their imagination run away with them. For example, the Italian Count Gianni Caproni, a pioneer of successful large bombers in World War I, was convinced he could achieve equal success immediately after that war with a fantastic civil transport aircraft called a *Noviplano*. This did not mean 'new plane' but 'nine-wing': Caproni proposed to use three sets of triplane wings, one near the nose, one half-way along the fuselage and the third at the tail! Any aerodynamicist could have told him that such an arrangement is not only hopelessly inefficient but bound to cause disastrous longitudinal instability. Caproni, however, could tell everyone in his company what to do, so the first *Noviplano*, the Ca 60, was actually built.

The span of each wing was a modest 98ft 5in, but their total area was an amazing 8,073 sq ft. At the front were four 400 horsepower Liberty engines, with another four at the back. Because Caproni considered his monster unsuitable for the tiny grass fields that then served as aerodromes, the Ca 60 was a flying-boat. Its fuselage (or, rather, hull) was 77ft long and in appearance combined the least attractive features of a giant railway coach and a houseboat. At the front was an open cockpit for the two pilots; the 100 passengers, however, were to be accommodated in the greatest luxury. The *Noviplano* was launched on Lake Maggiore on January 21, 1921. Well, at least it floated. But then came its maiden flight on March 4. The *Noviplano* did take off, but then had second thoughts: its inherent instability caused it to arch over and dive into the lake. The fiasco at least had the virtue of dissuading Caproni from pressing ahead with his greater vision of a 100-ton *Noviplano* to carry (or no doubt fail to carry) 150 passengers across the Atlantic.

It is wrong (as well as pointless) to discourage such visionaries, as long as their failures do not kill people or bring discredit on aviation. Another such was the German, Dr Hugo Junkers, who as early as 1909 – when Blériot was spluttering across the Channel in his 'stick-and-string' monoplane – was not only making drawings of gigantic all-wing aircraft to be made of metal but even taking out a patent. In World War I his company produced the world's first all-metal biplanes and monoplanes, bringing a new toughness to a species of vehicle which previously could be damaged by leaning a ladder against the fuselage. To make his aircraft even less prone to damage, the skin panels were all corrugated.

When World War I ended in November 1918 Junkers at

OO-AIS

once began drawing plans for huge civil aircraft, both landplanes and flying boats. None of these were built however, for the Allies would not permit it. But Junkers was allowed to build a prototype of the F 13, a neat six-passenger machine. The corrugated skin did not detract from its modern appearance and the low cantilever monoplane wing had no struts or bracing wires. It proved a worldwide success, and 322 were sold. It led to hundreds of other cantilever monoplanes, including two examples of the giant G 38 in 1929–32. This was a tangible expression of Junkers' wish to build an all-wing machine because, although it had a tiny fuselage to carry the tail, the whole machine was dominated by the gigantic wing; its passengers could sit in the wing, looking out through windows in its leading edge. Later, in the 1930s, the Ju 52/3m, a three-engined, 17-seater became the best-selling of all European airliners; total production (mainly for the Luftwaffe in World War II) amounting to 4,835.

During World War I the leading German constructor was actually a Dutchman, Anthony Fokker. After the Armistice he took many trainloads of aircraft parts and materials out of Germany, under the noses of Allied officials, and set up again in his own country. With his designer Reinhold Platz he perfected a method of construction in which the fuselage was made of welded steel tubes, covered in fabric, and the monoplane wing (mounted above the fuselage) was a thick cantilever structure of wood, with ply covering. Platz had designed the F.II before the war finished, and the first example flew in October 1919. It was quite a small machine, with an engine of 185 horsepower and seats for four passengers in the cabin with a fifth beside the pilot in the open cockpit. Many other Fokker airliners followed, the most successful of all being the F.VII. This like the Ju 52 was first produced as a single-engined machine and subsequently modified as the F.VII/3m with three engines of lower power. There were literally dozens of versions of this airplane, with many different kinds of engine. Some individual aircraft became famous, notably the *Josephine Ford* (the first of the three-engined version), which in May 1926 was the first airplane to fly over the North Pole; and the *Southern Cross*, flown by Sir Charles Kingsford Smith all over the world, including the first flight across the Pacific in May/June 1928. Both these aircraft have survived to this day.

Another German who dreamed of building giant airplanes was Dr Claude Dornier, and in his case the dreams became reality. In World War I he built a succession of enormous flying-boats characterized by their all-metal construction – then a radical feature. In the 1920s his Wal (Whale) was the most successful flying-boat in the world. The first, flown on July 31, 1919, was sunk by the occupying Allies in 1920; but Dornier evaded their restrictions by setting up a company in Italy. Well over 150 Wals were made there, as well as 40 in Spain, about 40 in Holland and three in Japan. Larger flying boats followed and on July 25, 1929 the stupendous Do X made its first flight. It was, by a comfortable margin, the biggest, heaviest and most powerful airplane ever built up to that time.

A corrugated-skin, all-metal monster, it had a flying-boat hull with a stub wing (called a sponson) low on each side, to give lateral stability when afloat. The hull resembled that of a ship, and on top rested the rather ungainly untapered wing. Above that again were the 12 engines in six back-to-back pairs. The wing span was 157ft 6in, and the loaded weight 123,459lb. The interior was furnished for 66 to 72 passengers, but on October 21, 1929 the Do X took off with a crew of ten, 150 passengers and nine stowaways – the total of 169 was not surpassed until long after World War II. This famous machine made a tour of the Americas, and two were delivered to Italy; but the Do X suffered from an extremely poor rate of climb and ceiling, and never had the range Dornier had planned. As a result none of the three built earned their keep in commercial service.

Back in 1922 Junkers, like Dornier, had evaded the Allied restrictions by setting up factories outside Germany. One was in Sweden, and a bigger one was at Fili, near Moscow. The Soviet Union wanted warplanes to help complete mopping-up operations after the civil war and suppress unrest in many places, and these were provided partly by the Junkers factory. The German method of all-metal construction also impressed a young Russian designer, Andrei Tupolev, who began designing a succession of ever-better cantilever monoplanes. Thanks to a most gifted team of engineers, Tupolev was soon producing world-beaters, such as the ANT-6 of 1930 (the TB-3 heavy bomber) and the ANT-16 (TB-4 heavy bomber), which was almost the biggest and heaviest machine of its day. Then came a request to create the flagship of the Maxim Gorky propaganda squadron. Hundreds of Soviet officials were involved in various committees, but the responsibility was Tupolev's alone.

The result, the ANT-20, first flew on June 17, 1934. Although it was lighter than the Do X, at 92,593lb, it had a greater span (206ft 8in) and greater power (eight 900 horsepower engines, six on the wing and two in a push/pull nacelle above the fuselage). It was a far better design than the Do X, and the interior was fascinating. Sometimes it was required to carry passengers – for example, as a reward to local commissars and other officials – and then it normally seated 72. Alternatively, it was used as a propaganda vehicle, with a crew of 20 to 23 and every conceivable kind of communications equipment, including radios, telephones, sky-shouting loudspeakers, arrays of electric lights under the huge wings for displaying slogans, photo libraries and film projectors, and a high-capacity printing press. Everyone was very proud of the ANT-20, which was popularly known as the *Maxim Gorky*, until on May 18, 1935 it was destroyed by a mid-air collision with an escorting fighter, which was performing unauthorized

***Above left*: In 1934 the Tupolev ANT-20 *Maxim Gorky* was the world's most powerful airplane and the biggest landplane. *Left:* This Fokker F.VII/3m was made in Belgium.**

The Dornier Do X was flown in 1929 with 12 air-cooled radial engines. It is seen here after being modified with 12-cylinder water-cooled engines.

aerobatics around it. An improved successor was built, the ANT-20bis, but this was used as a conventional hard-working civil transport until it was damaged beyond repair on December 14, 1942.

Planes for Everyman?

In 1919 there developed a wave of popular enthusiasm for the notion of the private-owner light airplane. Newspapers wrote of the prospect of an airplane in every garage. In Britain the Blackburn company designed a small monoplane seating pilot and passenger (or pupil) side-by-side. Called the Sidecar, it was to cruise at 83mph on a 40 horsepower engine, with a fuel consumption of 27 miles per gallon. The prototype was exhibited at Harrods of London in April 1919 and priced at $730 (£450); but even Blackburn never really took the prospect of mass production seriously. Then followed a shoal of light aircraft, many of them produced in response to competitions run by the Air Ministry and the London *Daily Mail* to try to pick the best airplane to be made for private owners and clubs, both for touring and for instruction and aerobatics. Competitions and trials were held from 1923 to 1925. They might have been useful and important if the officials who laid down the rules had refrained from stipulating various pointless numerical values, such as an upper limit of engine size of 750cc for single-seaters and 1,100cc for two-seaters. The result was a swarm of nice little airplanes which were completely useless for any practical purpose: they had neither the power to fly properly nor the structural strength and all-round toughness to stand up to years of flying, often by barely competent pilots.

Captain Geoffrey de Havilland and his design team decided to build what was needed, and they asked Major Frank Halford to produce an engine. The latter materialized as the Cirrus, with four air-cooled cylinders and giving 60 horsepower, about double the power of its many predecessors. The resulting airplane was the D.H. Moth, first flown by de Havilland on February 22, 1925. A simple biplane, with two cockpits in tandem, it had a ply-covered fuselage and fabric-skinned folding wings. To say it was a success is to damn it with faint praise. Moths went all over the world. By the end of 1928 over 400 had been delivered. In 1929 production started in the USA, some 120 being delivered in the first year. Soon the original engine gave way to engines of 75, 90 and 105 horsepower, followed from 1927 by the legendary D.H. Gipsy

DO-X
LUFT HANSA
D-1929

G-ALUN

The beautiful SR.45 Princess was the biggest airplane ever built in Western Europe. Sadly the three Princesses were completed just as flying boats became obsolete.

engine which grew from 85 to 140 horsepower. Many other species of Moth followed, including the Puss Moth and Leopard Moth cabin monoplanes, the Fox Moth – which, on 120 horsepower, carried a pilot and four passengers! – and, most famous of all, the Tiger Moth trainer, of which more than 8,000 had been built in many countries by the end of World War II.

In 1934, to mark the centenary of the Australian state of Victoria, a challenging air race was held between Mildenhall, near Cambridge, England and Melbourne. It was obvious that no British aircraft existed with anything like the speed and range needed to beat the modern American stressed-skin machines. Patriotic de Havilland said: 'Even if we lose money, we must offer to build a special aeroplane to win this race.' The result was the D.H. 88 Comet. Since the de Havilland company had no experience of all-metal, stressed-skin construction, the Comet had to be made of wood; but it was almost the first British aircraft devoid of struts or bracing wires. The wheels retracted into the nacelles behind the two engines, and the fuselage could store sufficient fuel for a range of almost 3,000 miles. Aft of the fuel tanks were the tandem cockpits for the two pilots who could share the flying. Three Comets were sold at once, and two more later, and one won the race in the outstanding overall time of 70hrs 54min. This aircraft was disposed of as scrap in 1936, was rescued, and then spent 13 years forgotten under tarpaulins. It was rediscovered in time to be displayed in the 1951 Festival of Britain, when holes were drilled through the wings so that it could be hung from above! More years of neglect followed, but in 1977 work began on what amounted to a total rebuild, and today this famous aircraft is back in the sky.

Enter the DC-3

Of all the competitors in the 1934 race the most dangerous to the Comet was a Douglas DC-2 of KLM, the Dutch airline. Unlike the Comet, this was a working airliner, and an example of modern American stressed-skin construction. It got to Melbourne only a little way behind the Comet racer, and won the handicap race because it carried a load of passengers and mail. Douglas had built the original DC-1 (DC stands for Douglas Commercial) for the airline TWA in 1933, and then it was slightly enlarged into the DC-2. Douglas was now on its way to massive sales all over the world, simply because the

The Hughes' 'Spruce Goose' made only one brief flight, but in many respects it is still the biggest airplane of all time.

DC-2 combined all the new technologies: powerful and efficient engines housed in low-drag cowlings and driving variable-pitch propellers; thin wings with no struts or wires and fitted with flaps; retractable landing gear; and a completely streamlined overall shape. While building an eventual 220 DC-2s, Douglas flew the first of a further enlarged version, the DST (Douglas Sleeper Transport) on December 17, 1935. While the DST was never important, the corresponding 'day plane', the DC-3, was to become the most famous transport airplane of all time. By the end of World War II more than 14,100 had been delivered in dozens of versions, including 2,930 built in the Soviet Union and 487 in (enemy) Japan. Early versions seated 21; wartime military models had strengthened floors for cargo, double doors and, often, inward-facing paratroop seats and a towing cleat for a glider. Post-war conversions often seated as many as 32. Hundreds are still flying in virtually every country in the world.

Aeronautical Dead-Ends

When the DC-3 was new, in 1936, airports were by modern standards tiny, and only a few had paved runways. A high proportion of the largest airliners were flying-boats, because longer take-off and landing runs were available on suitable stretches of sheltered water. Most of the trunk routes of Pan American and Imperial Airways, for example, were flown by flying-boats. These aircraft were mostly slower and shorter-ranged than equivalent landplanes; but Major Robert Mayo invented a Composite Aircraft consisting of a heavily loaded seaplane, unable to take off by itself, riding on top of a large but lightly loaded flying-boat. In the air the seaplane could then unlatch from its helper and go on its way. The seaplane, named *Mercury*, flew non-stop from Scotland to South West Africa in October 1938 to set a distance record for water-based aircraft of 5,997.5 miles, which stands to this day.

One reason it still stands is that today there are few seaplanes and flying-boats – though in World War II there were thousands. One of them was the biggest airplane ever built. In 1942 German submarines were sinking hundreds of thousands of tons of Allied ships each month. The American aviator and businessman Howard Hughes offered to build a colossal flying-boat, a veritable flying ship, which the dreaded U-boats would be unable to sink. At first partnered by shipbuilder

Henry J. Kaiser, Hughes designed and built the Hercules, an incredible machine made almost entirely of wood to avoid using strategic materials. No other aircraft has ever come near to equaling its wing span of 320ft. Its leading edges carried eight 3,000 horsepower Pratt & Whitney Wasp Major air-cooled radial engines driving four-blade propellers of 17ft 2in diameter. Under the bottom floor of the vast hull were 14 fuel tanks, each of 1,000 U.S. gallons. Above were floors stressed to take any vehicle in the U.S. Army, including at least two of the heaviest tanks, or alternatively 700 troops and their equipment. The sheer scale of the aircraft was without precedent, and it was no surprise when the war ended that the Hercules was still incomplete. But Hughes was determined that posterity should remember the monster as a success, not a failure, and at his own expense he finished it and flew it – just once – on November 2, 1947. Not the least remarkable part of the story is that this flying boat – popularly called 'The Spruce Goose' by the media – should still exist and be on public view at Long Beach, California, next to another ocean giant, the British liner *Queen Mary*.

Hughes never considered putting the huge flying-boat into any kind of commercial service. By 1947 the world was pretty well equipped with long paved runways, and faster and more efficient landplanes were replacing the few remaining flying-boats. Yet in 1952 Britain flew a new flying-boat which, although not quite as big as the Hercules, was actually heavier and much more powerful. This was the Saunders-Roe Princess, and it was the first of three ordered by the Ministry of Supply in 1946. It was the intention that they would be flown by BOAC – the predecessor of British Airways – non-stop between Southampton and New York, cruising at 380mph at 35,000ft and carrying 105 passengers in great luxury. Unlike the Hercules it was all-metal, the double-deck hull being one of the biggest pressurized fuselages ever built. Its weak spot was the powerplant. The Princess had been made possible by the development of turboprop engines, but the one chosen was short on power and very troublesome. Ten Bristol Proteus engines were arranged in four coupled pairs and, outboard, two single engines. All three Princesses were built, but the second and third were never flown and eventually, after endless arguments over its future, the first was scrapped in 1967. One reason for this sad outcome was the long time taken to develop and build the Princess in an age of rapid technical change. Another was the ridiculous procurement system which interposed the British government between the supplier and the operator. With a sensible system, either BOAC would have operated three Princesses, possibly with success, or they would never have been ordered in the first place.

Precisely the same thing happened with a huge British landplane. During World War II a committee chaired by Lord Brabazon decided what kinds of commercial transport Britain should build after the war. The first on the list, called the Brabazon, was to be an enormous machine with sufficient range to fly non-stop against the wind from London to New York. At that time this was almost impossible, and it would require a giant aircraft even to carry a small payload. The Brabazon was planned to carry 224 passengers, or 80 with sleeping accommodation. The contract was given to Bristol, and eventually the first Brabazon flew on September 4, 1949. In many ways it resembled a landplane edition of the Princess, although of course the pressurized fuselage was much slimmer and more streamlined. The engines were eight 2,500 horsepower Bristol Centaurus arranged inside the wing in four coupled pairs and driving four contra-rotating propellers on slim mountings projecting ahead of the leading edge. It was planned to complete the second as a Brabazon II, with four pairs of Proteus turboprops identical to those used in the Princess. But by 1954 it was evident that, using the latest technology, the route could be flown by a smaller aircraft, and that even the 330mph Mk II aircraft would be unable to compete against the soon to be available long-range jets.

The First Jet Airliner

In World War II the jet had revolutionized the performance of fighters and tactical bombers; but, because of its high fuel consumption, it had not been considered a practical proposition for long-range aircraft. So when the Brabazon committee recommended the development of a commercial jet, they thought in terms of modest range and small payload; and indeed the de Havilland company launched the design of the D.H. 106 Comet (no relation to their earlier Comet racer) as a mailplane of limited capability. At this time there was a strange belief that jet propulsion meant the adoption of unusual configurations, and the Comet was originally planned as a tailless machine with three engines at the back. Fortunately the designers at Hatfield had the vision to see that what the airlines would want would be a more conventional airliner able to carry a proper load of passengers. And when the Comet flew on July 27, 1949 it had the usual type of fuselage seating 36 in comfort, and with a normal tail at the back.

In many respects, however, the Comet was a glimpse of the future. Buried in the roots of the slightly swept wing were four D.H. Ghost turbojets, each of 5,000lb thrust. The wings were sealed to form integral tanks for 8,400 gallons of kerosene fuel. All flight controls were fully powered by Lockheed irreversible hydraulic rams. Because the Comet was aerodynamically so clean, it was fitted with airbrakes which could be extended under hydraulic power to permit fast let-downs at modest speeds. And because it doubled the cruising altitude of propeller-driven airliners from 20,000 to 40,000ft, the fuselage was pressurized to the unprecedented level of 8.25lb/sq in. The result was an aircraft offering a smooth ride at twice the speed (about 500mph) and devoid of fatiguing vibration, so that passengers arrived refreshed in half the time.

Services began on May 2, 1952, and it was soon evident that the Comet's enormous passenger appeal more than overcame a relatively short range and supposedly unimpressive economics. Airlines gradually joined the queue, while de Havilland rushed through the Comet 2 with 44 seats, more fuel and more powerful and efficient engines, and then the Comet 3, with more powerful engines still and a fuselage stretched to seat from 76 to over 100 passengers. This was assured of major orders from all over the world when, in 1954, two Comets of BOAC exploded in mid-air. It was eventually established that, owing to stress concentrations caused by making cut-outs in the fuselage square instead of round, the skin had begun to crack at the corners and eventually, as the result of fatigue, it ripped apart. It took three years to get a safe Comet 4 into the sky – and by that time the world markets had been won by the Americans.

Boeing had flown their prototype 707 in July 1954 and, thanks to enormous orders for tanker versions for the U.S. Air Force, they were able to develop larger models for the world's airlines. Douglas had to compete, and though they had no benefit of any military orders they initially level-pegged Boeing with the DC-8. Both were much bigger, faster and longer-ranged than the Comet, and they were powered by four turbojets of about 13,000lb thrust hung in pods well below and ahead of the sharply swept wings. Early versions had inadequate range to fly the North Atlantic. So did the Comet 4; but in a childish 'race' BOAC put the Comet 4 on the London to New York route on October 4, 1958 to beat Pan Am's first 707 service by a few days. The race proved little, because the 707 and DC-8 made a mockery of the many experts who predicted that buying jets would prove a financial disaster for the airlines. They absolutely transformed the air-transport industry, bringing new speed, comfort and reliability, (and incidentally for the first time freeing the industry as a whole from the need for subsidies). These aircraft did, however, involve the airport authorities of every country in enormous expenditure on lengthening and strengthening runways. But the improved airports were then ready for the even bigger and longer-ranged aircraft that were to come.

In 1955 SNCA du Sud-Est in France flew an attractive aircraft with a Comet-size fuselage but only two engines, both hung in a completely new place: on the sides of the rear fuselage, where their noise was remote from the passenger cabin. Named the Caravelle, it was the first short-haul jet to be sold to airlines, and despite the fact that turboprop airliners burned much less fuel, were generally much quieter at airports

***Right*: The D.H.60 Moth triggered off the great private flying movement around the world.**
***Inset*: Ten years later D.H. built the Comet racer.**

SS
G-ACSS
34
G-EBWD

G-AGPW

TWA
NC13711
TWA

and very nearly as fast, the short-haul jet soon caught on. As well as 280 Caravelles, airlines bought 117 Tridents, 230 One-Elevens, 1,832 Boeing 727s, over 2,000 DC-9s and MD-80s and over 2,500 Boeing 737s. This was despite the sudden, crippling rise in the price of fuel following the Middle East oil crisis in 1973.

In the 1970s various research teams intensified their interest in propfans, which are propellers with an exceptionally large number of blades, each much thinner than normal, sharp-edged and curved like a scimitar. Powered by a turboprop, a propfan can drive an airliner at jet speed yet with fuel efficiency similar to that of turboprops. It seemed for a time that propfans would soon sweep away both the jets and the traditional turboprops. Boeing offered the propfan 7J7 and McDonnell Douglas the MD-90. Meanwhile Airbus offered a traditional jet, the A320, but packed it with more new technology than had ever before been fitted into one airplane. The answer was clear: the more airlines studied the new technology of the A320 – most of it of an electronic nature designed to increase flight safety – the more they liked what they saw. Over 400 orders and options had been placed before the first A320 flew in February 1987, and by 1990 the total for the 320 and for the stretched 186-seat A321 had reached about 1,200. Observing this, Boeing put the 7J7 on ice and McDonnell Douglas launched the MD-90 with conventional jet engines (indeed, the same engine that is fitted to many A320s).

The engines are all high-ratio turbofans, which are almost like ducted propfans. Most of the thrust comes from a giant fan on the front, only a little being contributed by the hot core jet which drives the fan. The first of this new species of engine was a giant turbofan fitted to the Lockheed C-5A, an enormous airlift cargo aircraft for the USAF. Pratt & Whitney developed a similar engine and in 1966 this was picked by Boeing to power the 747. When this flew in 1969 it was dubbed by the media 'the Jumbo Jet', because it was the biggest thing ever seen at an airport. Early versions had four engines of 40,000lb thrust, weighed 710,000lb and could carry 366 passengers about 6,300 miles. Today the 747-400 has engines of about 60,000lb thrust, weighs 870,000lb and carries 412 passengers more than 8,400 miles. Whereas a feature of the original 747 was that all accommodation was on one main deck, the 747-400 has an upper deck at the same level as the flight deck. Just as the 707 pioneered the jet age around the world, the 747 pioneered the mass movement of people and cargo at prices which, in relation to inflation, have become ever-cheaper. Like so many of the world's greatest aircraft, the 747 at the outset seemed a frightening financial risk, the bill for development and starting production being many times greater than the net worth of the company. At that time the only order was Pan Am's launch buy of 25, at about $35 million each. Today a 747-400 costs about $150 million, and the order-book is about to pass the 1,000 mark.

Apart from the C-5 cargo aircraft, which is fractionally larger though less powerful, the only airplanes in the sky bigger than the 747 are the gigantic cargo airlifters designed at the bureau named for O.K. Antonov at Kiev, in the Ukraine. Antonov had produced a succession of giant turboprop aircraft, most of them cargo aircraft, when in 1965 he startled the aviation world with the An-22 Antheus. This had a span of 211ft 4in, and because the wing was of relatively small area the wing loading was very high – 148.5lb/sq ft at the maximum weight of 551,160lb. Powered by four of the mighty NK-12MA turboprops of 15,000 horsepower each, the An-22 could cruise at 400mph carrying a load of 80 tons for 3,100 miles.

Ten years later Antonov was working on the An-72 and An-74, which, although much smaller, use advanced flap-blowing and reverse thrust to operate from short airstrips. The An-74 is specially designed for cargo transport in polar regions. While the An-72 and 74 were being readied for production, work went ahead on the next generation beyond the An-22: the mighty An-124 Ruslan, named for a Russian folk hero. Unlike the An-22, the An-124 is a jet, very much in the class of the American C-5 but even bigger and far more powerful. The engines are four Lotarev D-18T turbofans of 51,590lb thrust each. They are hung from a swept wing of just over 240ft span, mounted high on a vast fuselage offering an unobstructed cargo hold more than 118ft long, with loading doors and ramps at each end, the height and width being 14ft 6in and 21ft respectively. The main landing gears consist of 10 wheels on each side arranged in a row of five twin-wheel units retracting inwards. Maximum weight of the An-124 is 892,872lb, and a cargo load of 150 tons can be carried almost 3,000 miles.

The An-124 was impressive enough, so nobody was quite prepared for the monster the Antonov bureau brought to the 1989 Paris airshow, with the Soviet Buran space orbiter riding on its back. First flown on December 21, 1988, the An-225 Mriya (the name is Ukrainian for 'dream') is simply an An-124 stretched in all directions. The wing was extended to a span of 290ft by adding an extra inner section with another pair of engines. The fuselage was lengthened to give a cargo hold over 141ft long, overall length being nearly 276ft. Two more pairs of landing wheels were added on each side, the rear four pairs on each side, and the two twin-wheel nose units, all being steerable. To facilitate the carriage of giant 'pick-a-back' loads the tail was redesigned with twin fins and rudders. There are rest areas for a relief crew, and at the rear above the cargo hold is a relatively tiny area where seats for 70 passengers can be installed. Maximum weight of this colossus is 1,322,750lb, and it can carry a 250-ton payload.

Chief designer Pyotr Balabuyev was asked how many Mriyas might be built. He said 'There will be others, but not very many. And so far as we can see, this is about as big as we can get. Anything larger simply wouldn't suit the world's airports.'

***Above left*: The mighty Brabazon.**

***Left*: Not nearly as impressive, but far more successful, was the DC-2.**

The world's first jet airliner, the D.H.106 Comet, in July 1949. It still looks attractive over 40 years later.

The Boeing 707 was the first of the 'Big Jets' to transform air travel throughout the world from the mid-1950s.

BOEING
707

Called the Jumbo Jet, or even the The Aluminum Overcast, the Boeing 747 revolutionized air travel every bit as much as its ancestor, the 707.

PAN AM

Today designers are dreaming of giant 'Spanloaders' to cut the cost of air cargo. Huge freight containers are carried inside the wing space and loaded through the hinged tip.

First flown in 1954, the Lockheed C-130 Hercules is still in production today.

The old 'Spruce Goose' still has the biggest wingspan but the An-225 Mriya is the biggest, heaviest and most powerful airplane of all time.

4

INTO THE UNKNOWN

This aerospace plane, proposed by the West German company MBB, could carry passengers and cargo to the other side of the world in about 2 hours, operating from ordinary airports. It is named Sanger after the Austrian rocket pioneer of the 1930s.

INTO THE UNKNOWN

MAN IS A LAND-BASED, AIR-BREATHING ANIMAL. IT IS NO more natural for him to fly than it would be to live under water. So every significant advance in the history of aviation has been a step – often a perilous one – into the unknown. Over the past 200 years many of the more reckless would-be aviators would not have attempted to fly their bizarre contraptions had they been able to foresee the humiliating or lethal outcome. Today, with our vast accumulation of knowledge about the Earth's atmosphere, about aeronautics, and about engines and materials' technology, the dangers of each step forward may be more predictable; but aviation has as great a need as ever for the brave and questing spirit of the pioneers.

Above: **The Rutan long-EZ (Long-easy).**
Right: **The Rutan globe-girdling Voyager.**

Of course, the more prudent pioneers attempted to avoid taking unnecessary risks, however small. The Paris-based Brazilian Alberto Santos-Dumont, who flew his airship around the Eiffel Tower in 1901 and five years later made the first flight in Europe in a heavier-than-air machine, gave much thought to the future of civil aviation. One of his more arcane pieces of research in this field was to eat his lunch while suspended 10 feet in the air, the better to prepare himself for the day – in the not too distant future, he was sure – when in-flight meals would be served aboard airliners and airships. (His prescience was better than the quality of his research: it was only a few years later, in 1910, that the Delag company began serving gourmet dinners aboard their Zeppelin airships flying over the major cities of Germany.)

The Long-Distance Heroes

For many aviators, the only spur needed has been the chance to be a winner – to fly faster, farther, higher or longer than anyone before them, or to be the first to achieve a difficult or dangerous feat. The history of aviation between the two world wars is full of the deeds of such men. That phase began on the morning of June 14, 1919, when two RAF officers, John Alcock and Arthur Whitten-Brown, took off in their twin-engined Vickers Vimy bomber from a field at St John's, Newfoundland, heading east. Exactly 16 hours 28 minutes and 1,890 miles later they landed in a bog in western Ireland, having made the first direct crossing of the North Atlantic. Their fast time – they averaged about 114mph – was achieved in spite of almost overwhelming difficulties, which more than once threatened to cause the Vimy to plunge into the ocean. Many other bold aviators faced similar dangers. Probably the most famous pilot of all time was Charles Lindbergh. At 07.52hrs on May 20, 1927 he opened the throttle of the Wright Whirlwind engine of his little Ryan monoplane and watched the bumpy grass of New York's Roosevelt Field begin to move with excruciating slowness past his side window – his only view of the outside world in the absence of a windshield. Grossly overloaded with fuel, the Ryan trundled almost the whole length of the field until, after a series of hops and bumps, it stayed in the air and just cleared telephone wires at the end. The little Ryan was now set to face the unknown dangers of 3,610 miles of ocean. Lindbergh had had no sleep the previous night and he was to have to force his tired eyelids to stay open for another 33 hours 39 minutes. He wrote: 'I let

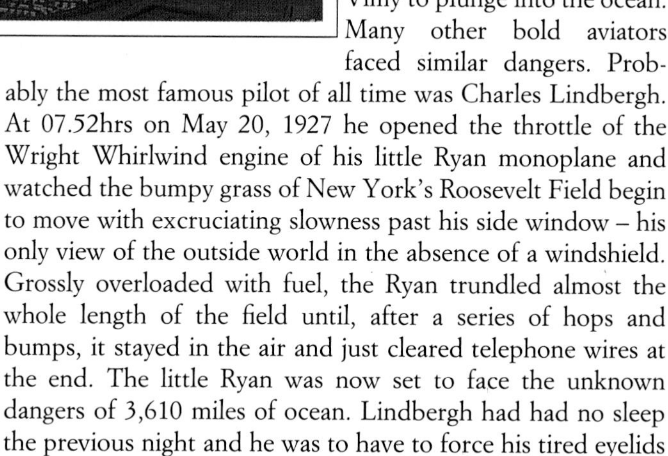

N269VA

CRIZ

S
1595

my eyelids fall shut for five seconds; then raise them against tons of weight. Protesting, they won't open wide until I force them with my thumb, and lift the muscles of my forehead to keep them in place.' Strangely, as he neared the end of his journey, Lindbergh returned to full wakefulness. He crossed the Irish coast only three miles off his intended track, and eventually made an amazing night landing (remember, he had no forward view whatever) at Le Bourget airport to be met by half the population of Paris.

Many other aviators have set forth into the unknown and triumphed over time, distance, and fear. One of the most remarkable feats was the first solo flight around the world. It was made in a single-engined Lockheed Vega by a one-eyed pilot, Wiley Post, between July 15 and 22, 1933. He had to keep his one eye open not just for 33 hours, like Lindbergh, but for nearly 7 days 19 hours. It is difficult to think of a comparable test of endurance, but two Americans managed it in 1986 with a feat that is impossible to follow. Richard G. (Dick) Rutan, brother of aircraft designer Burt Rutan, teamed up with Jeana Yeager to create an airplane able to fly around the world non-stop. The Voyager looked rather like a beautiful sailplane, with a structure mainly of graphite fiber. The 110ft 10in wing sagged on the ground under the weight of gallons of fuel, but curved upward in flight. The tail was carried on twin booms, linked by a foreplane to the central nacelle with two Teledyne Continental piston engines. At the front was a 130 horsepower aircooled O-240 and at the back a pusher 110 horsepower liquid-cooled Voyager 200. In between there was just room for Rutan and Yeager.

A few seconds before 08.00hrs on the morning of December 14, 1986 Rutan opened up both throttles fully. Ahead stretched the vast expanse of Edwards Air Force Base – and the world. At first the fuel-filled wings rubbed on the ground, grinding away the tips with their winglets which were later deliberately shaken off. At last, after two minutes of a nail-biting snail-like acceleration and a run of 14,200ft, the strange bird reluctantly parted from Mother Earth and set course westward towards the Pacific. Eventually, as planned, the front engine was stopped and its propeller feathered, the flight continuing with power only from the small engine at the rear. At about 08.05 hrs on the morning of December 23, Voyager was back over Edwards. In just three minutes over nine days it had flown around the world non-stop, a distance of 26,678 miles. Until that time there had always been two world records for range: one for straight-line distance and one for a closed circuit. Rutan and Yeager had the unique distinction of breaking both records in one flight.

Above left: This Rutan is the Model 72 Grizzly, built to explore the STOL (short take-off and landing) qualities of tandem-wing airplanes.
Left: S1595 was the actual S.6B seaplane used by F/Lt John Boothman to win the Schneider Trophy for the third consecutive time in 1931.

The Quest for Speed

In the early days of steam railways, in 1825–30, it was widely publicized as an unassailable medical fact that, if a train was unwise enough to exceed a speed of 25mph, passengers' lungs would be sucked inside out. By the time people could fly in airplanes, trains had already nudged 100mph. This was twice as fast as the early flying machines, but by 1912 special racers were going beyond this figure, and by the start of World War I the record stood at 126.64mph. In 1921 the record passed 200mph; 300mph was reached in 1927, and 400mph in 1931.

The incentive to break the 300 and 400mph targets was the Schneider Trophy. This silver trophy was donated by the arms manufacturer Jacques Schneider in 1913 specifically for marine aircraft, which were to race a number of laps around a (usually triangular) circuit. Gradually the race became a matter of national honor and big business, and it spurred the development of seaplanes to a remarkable degree. By far the most important effect was on engine design. Schneider stipulated that anyone winning three races consecutively should keep the trophy. Britain just beat Italy to win the 1927 contest. For the 1929 race Rolls-Royce developed a special racing engine, known only as the R, which gave more power (1,900 horsepower) for less weight than any previous engine of any kind. It enabled a Supermarine S.6 seaplane to stave off the Italians and win again. Not to try for a third win seemed unthinkable, but in 1931 Britain was in the middle of the Depression and the government announced it would not enter that year's race. More or less at the eleventh hour Lady Houston, a well-known aviation enthusiast, undertook to pay for a British entry. Rolls-Royce toiled night and day in the few weeks remaining to wring ever more power out of a batch of largely redesigned R engines, and by the day of the race they were confident of getting 2,350 horsepower. Installed in an improved S.6B seaplane, it won the race at 340.08mph; and later in 1931 another R, running on a specially blended 'cocktail' of fuels, pushed the record up to 407.5mph.

Little did anyone know at the time that Lady Houston's patriotic gesture would prove to be significant in the longer term. The S.6B helped its designer, Reginald Mitchell (who was dying of cancer), to create the Supermarine Spitfire fighter; and the R helped Rolls-Royce from 1933 onwards to develop the Merlin engine, which powered the Spitfire as well as its more numerous partner in the Battle of Britain, the Hawker Hurricane. Without these two fighters Britain would almost certainly have been conquered by Nazi Germany.

Mach 1 and Beyond

Frank Whittle's invention of the turbojet in 1929 and how it removed the old limitations on the speed of airplanes has already been discussed. Putting Whittle's invention to use, however, was held up for seven and a half years. It was during this time that German turbojets and rockets made the design of remarkable new fighter aircraft in Germany possible. Moreover, unlike the Allies, the Germans did their utmost to exploit

20003
U.S. AIR FORCE

The amazing research airplanes that pioneered supersonic flight in the USA had to be dropped at high altitude from under a parent airplane, such as the B-52 seen here with an X-15 under the wing. The technique was started in 1946 with the Bell X-1 (inset far left), the first supersonic airplane. The X-15, seen landing (inset below) reached over Mach 6.

new aerodynamic concepts such as sweptback wings. In 1943 rumors of amazing new German jet and rocket aircraft caused the British government to issue a specification for the world's first supersonic aircraft. The Miles M.52 was urgently designed to fly on the level at 1,000mph, powered by a Whittle W.2/700 turbojet boosted by an afterburner. Pilot Ken Waller got ready to fly the bullet-like machine, which was nearing completion in 1946. Then the unbelievable happened: the post-war government canceled the M.52 at the stroke of a pen. It offered various reasons, such as the fact that it was dangerous, and insisted that 'the idea supersonic flight is just around the corner is totally erroneous'. Miles was instructed to hand everything over to Bell Aircraft in the United States.

In fact, Bell didn't need any help from Miles: they were about to prove that supersonic flight *was* just around the corner. The Bell XS-1, later called the X-1, made its first gliding flight on January 25, 1946, even before the British aircraft had been canceled. Then the XS-1 was taken to Edwards Air Force Base, a huge and remote testing site in California, where it was slung underneath the fuselage of a B-29 carrier aircraft. On October 14, 1947, it was dropped at high altitude from the B-29. Pilot 'Chuck' Yeager progressively fired three of the four combustion chambers of the XS-1's liquid-propellant rocket engine. Flying level at 40,000ft, the orange research aircraft quickly accelerated to Mach 1.06 – the first time anyone had traveled faster than sound. Later the same aircraft reached Mach 1.45 (957mph). Today the name of Yeager is famous all over the world; thanks to British shortsightedness hardly anyone has heard of Ken Waller.

Later versions of the X-1 had many design improvements; among other things they carried more rocket propellant, so they could be held at full power for longer, and the speeds kept increasing. This particular family of aircraft reached its peak speed on December 12, 1953. Yeager was again the pilot, and he took the X-1A to Mach 2.44 (1,650mph) at a little above 70,000ft altitude. But just before reaching this speed he attempted to correct unwanted rolling maneuvers, first to the left, then to the right. Suddenly the speeding aircraft went completely out of control. It not only rolled: it tumbled end-over-end. Yeager was subjected to such violent maneuvers that he lost consciousness. When he came to, the X-1A was in a spin, upside down, and had dropped no less than 36,000ft.

It says much for the toughness of man and machine that he

***Above:* The Fairey FD.2 beat the world air speed record by 310 mph, reaching 1132 mph.**

made a normal landing. What had happened had in fact long been predicted. It is called inertia coupling, and it stems from having too much mass in a long fuselage and insufficient fin area at the rear to give 'weathercock stability'. Many other aircraft were to suffer from it – some of them breaking up in the air – before designers knew precisely how to cure it.

Bell went on to produce the X-2, with a more powerful rocket engine, and on September 27, 1956 an X-2 reached Mach 3.196 (2,094mph). Meanwhile, the official world speed record stood at 822mph. By this time backward Britain had built a small supersonic research aircraft, the Fairey FD.2, and Fairey test pilot Peter Twiss urged the management to have a go at the world record. The FD.2's owners, the Ministry of Supply, declined to play ball and insisted that Fairey should pay the costs of the record attempt. Even Rolls-Royce, makers of the Avon turbojet, were extremely reluctant to participate. But Twiss eventually won through, and in a masterful display of precision flying (which involved figuratively 'threading the eye of a small needle' twice in each direction more than five miles above the ground) he set a new speed record on March 10, 1956, raising it to 1,132mph; indeed, if the FD.2 had had more fuel it could have set a record at about 1,400mph.

During this period Douglas Aircraft built their beautiful high-speed Skyrocket research aircraft for the U.S. Navy. In August 1953 one of them reached 83,235ft, and in November the Skyrocket became the first aircraft to exceed twice the speed of sound. Douglas's great designer Ed Heinemann then worked on the next step, the D-558-III. This was to have truly breathtaking performance. Its powerful rocket engine was to burn for 75 seconds, with the aircraft in a steep climb. Burnout was to occur at about 700,000ft, beyond which point the D-558-III was to dive back into the atmosphere, reaching (it was calculated) Mach 9 (6,000mph). Unfortunately the U.S. Navy declined to fund such probing into the unknown, so the next major advance was made by the X-15, a product of North American Aviation. This was rather bigger than the Skyrocket and did not aim quite so high (in all senses).

The first of three X-15s made its first gliding flight on June 8, 1959 and the first under the power of an interim powerplant (twin rocket engines similar to the single engine of the X-1 and Skyrocket series) on January 23, 1960. The first flight with the much more powerful XLR99 engine was on November 15, 1960. Altogether the X-15s made no fewer than 199 flights, some hair-raising and one fatal. The sum total of the information they provided is incalculable. No two missions were alike: different instrumentation and completely dissimilar objectives made each flight a fresh challenge. At first painted black, the X-15s were later rebuilt as X-15A-2s with far greater propellant capacity (their weight increasing from 31,275 to 56,130lb) and were covered with a thick layer of white ablative material to protect them against fierce thermal heating. Two of the records set were a speed of Mach 6.70 (4,520mph) and an altitude of 354,200ft (about 67 miles).

Jet Power and VTOL

In complete contrast, the advent of jet engines and turboprops made possible a totally new species of airplane, the VTOL (vertical takeoff and landing). A helicopter is also a VTOL but it is limited to low speeds, almost always much less than 200mph. The new kinds of VTOL, however, promised jet speeds; but there were many problems, mostly concerned with how to control the beast at very low speeds, at which normal control surfaces have no effect. One of the first experiments was conducted by Ryan for the U.S. Navy in 1948-51. An Allison J33 turbojet was upended and run in a vertical rig to see how well its rise and fall could be controlled merely by opening

***Right:* Startling to watch, but difficult to fly, the Convair XFY-1 – an early VTOL fighter. Equally difficult to fly was France's *Coléoptère* (inset) a VTOL jet.**

NAVY
NAVY
CONVAIR

Germany's VJ 101C seemed attractive, because its four wingtip engines could swivel bodily to give lift or thrust.

and closing the throttle. On May 31, 1951 it was hovering, secured only by loose tethers. Then a pilot's seat and controls were added, and on November 24, 1953 Pete Girard made the world's first manned jet-VTOL hovering flight. On May 28, 1956, Girard hovered in the Ryan X-13 delta-winged VTOL aircraft, which had previously flown as a conventional airplane. The big hurdle was reached on November 28, 1956, when an X-13 made the first transition in history, slowing down from high-speed flight to hover and then settle gently on the ground. Later, complete VTOL flights were made, starting with an accelerating transition and ending with a decelerating transition back to a VL (vertical landing).

But Ryan had competitors. Rolls-Royce had studied jet lift in World War II, and on July 3, 1953 Capt. R.T. Shepherd rode aloft on the thunderous thrust of two Nene turbojets built into a device aptly dubbed the 'Flying Bedstead'. Rolls-Royce now designed special lightweight jet engines which could be installed in large groups to lift any kind of airplane. A battery of four was fitted to the Short SC.1, which flew as a conventional airplane on April 2, 1957, lifted only by its delta (triangular) wing, and began hovering trials on May 23, 1958. SNECMA in France conducted similar trials with the Atar Volant, in which an Atar turbojet was mounted vertically with an ejection seat and controls on top; it was first flown untethered on May 14, 1957. The aim was to fit the Atar Volant inside an annular wing to create a *coléoptère* of the kind long dreamed of by Professor von Zborowski. The C.450 Coléoptère began flight trials on May 6, 1959, but it later crashed (the pilot ejecting successfully).

The sheer difficulty of piloting these early VTOLs was a major problem. It killed off two otherwise interesting designs for a small VTOL fighter powered by a powerful turboprop. Although this restricted top speed, the large contra-rotating propeller was far more effective than a jet in hovering flight. The Convair XFY-1 and Lockheed XFV-1 were extensively tested in 1954, both in conventional horizontal flight and in vertical flight, but work on both was abandoned in 1955. The difficulties had proved too great, especially in the decelerating transition and vertical landing, with the pilot lying head-down and looking back over his shoulder, and with no direct control over the rate of descent.

Continued on page 124

VJ101C-X1

The future of the Bell/Boeing V-22 Osprey tilt-rotor program is in jeopardy. Tilt-rotors can do anything a helicopter can do, and fly at two or three times the speed, and for greater distances, but funding for further development has been withdrawn.

eing

On the other hand, the need to create an air force that could be dispersed away from known airfields, and thus could survive in actual warfare, was obvious and urgent (although today most air forces simply ignore this need). The answer was the 'flat riser', the jet VTOL able to take off and land in the normal attitude (that is, pointing forwards, not upward). The pioneer flat riser, quickly assembled using parts of existing aircraft, was the Bell VTO, which first hovered on November 16, 1954. It had two J44 turbojets mounted on pivots to give either lift or thrust, and a control system using reaction jets at the wingtips and tail fed with compressed air from a small gas turbine amidships. With this experience Bell built the X-14, powered by two small turbojets with nozzles which could be vectored (swivelled) downwards or to the rear. This hovered on February 19, 1957 and made its first transition on May 24, 1959.

By 1960 jet VTOL was all the rage, with many companies planning to build prototypes. Only one, however, was to lead to a practical aircraft, and this was the smallest and simplest of all. The Hawker P.1127, powered by a forerunner of the Rolls-Royce Pegasus, first hovered on October 21, 1960. A long process of development followed, during which the thrust of the Pegasus engine was almost doubled to 21,500lb, before the Harrier GR.1 entered squadron service with the RAF at the start of 1969. Since then the obvious advantages of a survivable warplane have very gradually become appreciated, and the Harrier has undergone steady development.

Several of the prototype jet-lift aircraft flown in the 1960s were supersonic. Many arrangements of engines were tried, the German VJ 101C for example having six Rolls-Royce RB.145 engines, two mounted almost vertically in the fuselage and a pair installed on each wingtip in a nacelle that could be vectored to any angle from vertical to horizontal. One VJ 101C reached Mach 1.04. But then in 1966 a rival French machine, the Dassault Mirage IIIV, with eight RB.162 lift jets and a SNECMA TF106 propulsion engine, reached Mach 2.04. Today thousands of designers in the United States, Britain and Canada are trying to decide how best to create an ASTOVL (advanced short takeoff and vertical landing) aircraft. The work is being done at a leisurely pace, with the objective of getting something into service in the first decade of the 21st century. The only simple and proven system, the single vectored engine, has already been rejected as nothing like complicated enough.

All jet-lift aircraft can reach high forward speeds, but they are inefficient in hovering flight; they are also so noisy that the idea of 'stealth' becomes a joke. Conversely, a helicopter is efficient in hovering flight but is incapable of flying at jet speeds; and it is also quite noisy. In 1954, in an attempt to get the best of both worlds, Bell built the XV-3. It was the first tilt-rotor aircraft. At take-off it was almost a helicopter, being lifted by two rotors on the tips of the wings. Once well clear of the ground the pilot could begin the forward transition by gradually tilting the rotors forwards. This had the effect of accelerating the machine forwards, so that the wing could progressively take over from the rotors, which ended up as conventional, if very large, propellers providing forward thrust. The XV-3 was not taken further, but in the 1970s Bell began designing a completely modernized version, the XV-15. This was so successful that Bell and Boeing jointly designed a much bigger tilt-rotor, the V-22 Osprey.

First flown on March 19, 1989, the Osprey is a multi-role transport powered by two 6,150 horsepower Allison T406 engines. These are mounted in nacelles on the tips of the 46ft wing, which can be pivoted anywhere from upright to horizontal. Each engine drives a 38ft propeller/rotor with three giant blades made of graphite and glassfiber composites. After a vertical landing the Osprey can rotate its engines to the horizontal, fold the blades, and then turn the wing through 90 degrees to form a compact box-like package for parking on a ship. The Osprey fuselage is unpressurized because it has to serve many roles (an airline version might be pressurized and have a circular cross-section). The U.S. Marines want 552 of an assault version to carry 24 combat-equipped troops; the Navy wants 50 of a sea-rescue version and is also interested in up to 300 of an anti-submarine version packed with sensors and weapons; the USAF has a requirement for 55 of a special-forces transport version.

No helicopter could come anywhere near the Osprey in performance. It combines the ability to hover with a cruising speed of up to 345mph and a range of up to 2,400 miles because in cruising flight it behaves as an airplane. This new technology is certain in due course to take over at least half the market at present served by helicopters, and the lead held by the United States could be worth billions of dollars over the next 25 years. And yet powerful politicians in Washington have for two years been doing their utmost to kill the Osprey program. One of their arguments is that the missions could be flown by helicopters. So they could, if you need to travel only one-third the distance at less than half the speed. The potential wreckers appear to be unable to visualize the long-term importance of an environmentally acceptable vehicle that could link city centers, offices, factories or hospital pads at 350mph.

Advanced Technology Aircraft

Notwithstanding the wonderful and farsighted work done in the many giant laboratories of the U.S. National Aeronautics and Space Administration (NASA), most of tomorrow's aviation technology is funded to support military programs. Western aerospace companies are keenly watching to see what effects the East-West *rapprochement* and the succession of revolutions in the countries of the Warsaw Pact will have on the massive defense funding they have enjoyed for many years.

The biggest and most expensive programs in the West's

This Northrop XB-35 bomber first flew in 1946.

Seen here with landing gear extended, the XB-70 was, in 1964, almost the biggest, heaviest, most powerful, fastest, longest-ranged and noisiest airplane in the world. Built for the U.S. Air Force by North American Aviation, it was a reconnaissance bomber able to cruise for hours at Mach 3.

aviation industries are the ATB (Advanced Technology Bomber), the ATF (Advanced Tactical Fighter), and the ATA (Advanced Tactical Aircraft). The ATB is the Northrop B-2. The ATF has not yet been flown, though prototypes were due to appear in mid-1990. It is still a competitive program between two aircraft, the YF-22A being developed by Lockheed assisted by Boeing and General Dynamics, and the YF-23A being developed by Northrop, with McDonnell Douglas as principal sub-contractor. The ATA is the A-12, being developed in an equal joint program by General Dynamics and McDonnell Douglas.

One of the key meanings of 'Advanced' in the title of all three programs is that the aircraft are LO (low-observables) designs, popularly called 'stealth aircraft'. It is sobering to recall that in 1935 Sir Robert Watson Watt, the radar pioneer, should have pointed designers in the right direction, only to have the whole subject ignored for more than 40 years. Now at last, in the United States at least, stealth designs are mandatory for all new military and naval aircraft, although this enormously increases their cost. While they are not significantly bigger or heavier than the aircraft they are intended to replace, the ATF and ATA will probably have a unit price getting on for five times as great. This will sharply accentuate the trend towards fewer aircraft, each capable of a wider range of missions.

In the case of the ATF the difficulties are compounded by the fact that the chosen aircraft will be supersonic. Indeed, it will be a so-called 'supercruise' aircraft, able to fly a large part of its mission faster than sound without using its engine afterburners. Even so, every supersonic aircraft has to be very carefully shaped to minimize supersonic drag, and it is almost impossible to reconcile this with the need for minimum observability. By not using afterburners the temperature of the jets is greatly reduced; even so they would inevitably stand out like a sore thumb to any infra-red detector tuned to focus on very hot gas, though the hot nozzles of the engines can be hidden behind cooler parts of the airframe. As for their intense noise, this might not appear a serious problem because at high altitude there would be no-one in the open air to listen and at low level the ATF would precede its noise. The ATF, whichever one is chosen, will probably have canard foreplanes and engine nozzles able to vector the thrust up or down, partly for STOL performance and partly to enhance maneuverability in flight.

First flown in 1964, the Lockheed SR-71 (popularly called the Blackbird) was the fastest and highest-flying airplane ever to go into service anywhere. Withdrawn in 1989, there are no successors.

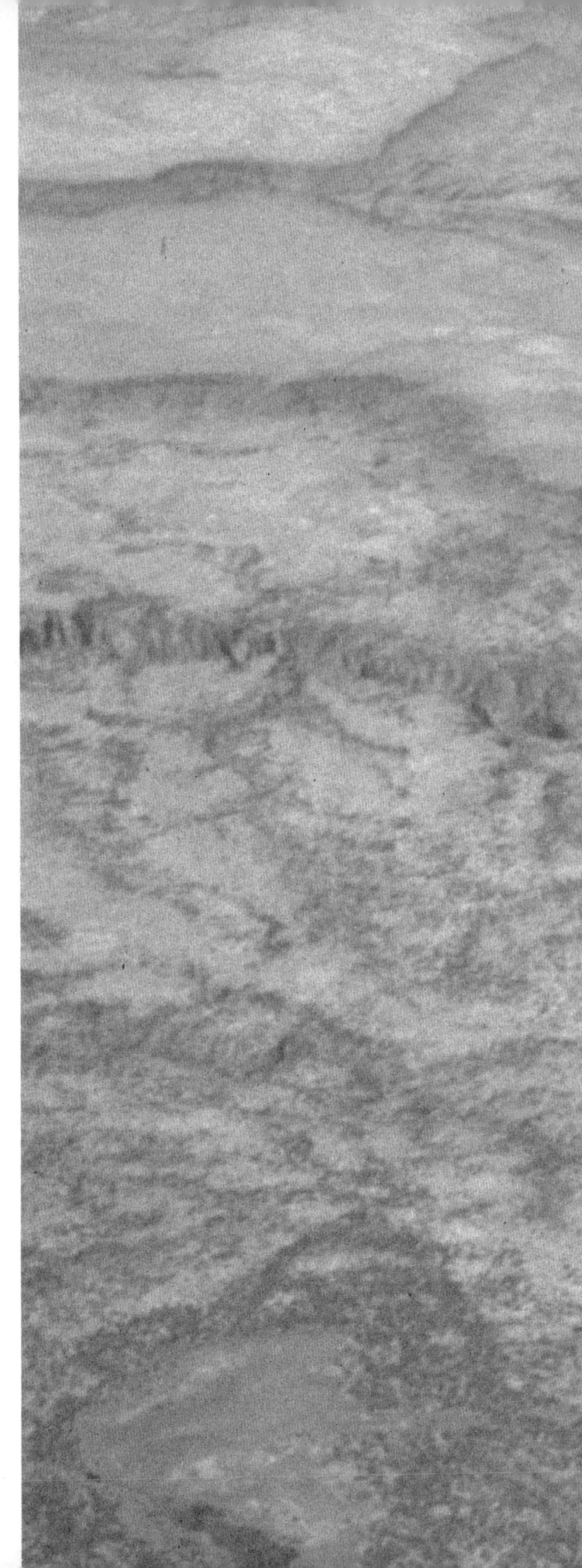

The A-12 Advanced Tactical Aircraft will be a carrier-based bomber for the U.S. Navy, though the U.S. Air Force is considering a land-based version as a possible replacement for the F-111. It will be powered by two turbofans derived from the General Electric F404. This engine is used with afterburners in the supersonic F/A-18 Hornet, and without afterburners in the first stealth aircraft, the Lockheed F-117A. Nothing has yet been disclosed as to whether or not the A-12 will be supersonic, but on balance it seems unlikely. Like the Grumman A-6F it is designed to replace, a subsonic A-12 would probably be less than half as expensive as a supersonic version, and it would carry much heavier ordnance loads over greater distances, besides being much more difficult to detect and shoot down. What is certain is that the A-12 will not be just an airplane on to which various items of sophisticated equipment are then bolted. The entire vehicle is being designed as an integrated system. Digital databuses will link every item in the aircraft, rather like a nation's railway network but of greater complexity. For instance, the multi-function radar will 'see' through a futuristic antenna system which almost certainly will involve hundreds or even thousands of self-contained transmit/receive modules covering large parts of the aircraft like the cells in a honeycomb. Similar modules, 'looking' in all directions, will warn of hostile radars or of missile attack and send out computer-controlled countermeasure signals aimed precisely at each threat.

Although the challenges of warplanes may be more exciting, commercial transports also keep the designers looking far into the future. For example, many engineers are still trying to perfect an all-wing aircraft. As mentioned earlier, in 1909 Dr Hugo Junkers in Germany dreamed of an all-wing airplane. It seemed obvious that, by doing away with everything except the wing, both weight and drag would be reduced, provided control could be retained. He was never quite able to realize this dream, though in the giant G 38 airliner of 1929 he came close to it. But in that same year the American Jack Northrop, a pioneer of all-metal stressed-skin construction, flew an all-wing airplane which he hoped would lead to a wholly new aircraft shape.

In 1940 Northrop flew an improved all-wing machine, followed in 1942 by the first of four N-9Ms, which were scale models of a giant bomber. The bomber itself, the XB-35, flew on June 25, 1946. A giant of its day, with a span of 172ft and

weighing 209,000lb (equivalent to almost four B-17s or Lancasters), it almost made it into production for Strategic Air Command. Later versions included the YB-49, with eight turbojets, and the YRB-49 with six. All looked like something straight out of the 21st century and, by sheer chance, their wingspan of 172 feet happens to be exactly the same as that of today's all-wing Northrop B-2.

In general, all-wing aircraft have not been successful, and in the case of the B-2 the configuration must have been adopted to minimize radar and visual signatures. But another class of all-wing (or almost all-wing) aircraft is the heavy cargo transports called spanloaders. All the giant U.S. builders of commercial transports have made studies for aircraft designed to carry standard freight containers of the kind seen on trucks, rail flatcars and ships. These would be loaded into a huge wing, either through a fold-up tip or through doors in either the tip or the leading edge. Many arrangements have been studied, most of them having a small tail for good stability and control in flight. Many experts have predicted that air cargo will increase enormously in the 21st century and that huge spanloader aircraft will be needed to shift it. Some people even believe that cargo airships are a good idea. Airship Industries in Britain and Westinghouse in the United States are leaders in airship technology.

Another focus of futuristic technology has for many years been Lockheed's ADP (Advanced Development Projects) center, popularly called the Skunk Works. Here a remarkable succession of unique aircraft has been created. One of the earlier ones was the Lockheed U-2. In 1955 it was announced as a 'research aircraft' for investigating the upper atmosphere; but this story shed all plausibility when, on May 1, 1960, a U-2 piloted by Gary Powers was shot down over Sverdlovsk, in the heart of the Soviet Union. It turned out that the U-2 had been designed to make clandestine overflights of unfriendly countries, using enormous cameras and other sensors to bring back imagery of the highest quality. To fly the mission, something resembling a jet-propelled sailplane was created, with a wing of enormous span and an airframe of minimum weight (which of course implied lower than normal strength). Over the years the U-2 was developed in some 30 different versions, including the greatly enlarged U-2R and related types; the final versions, designated TR-1, were intended for flying along the eastern borders of NATO territory in Europe while looking far into the adjoining countries of the Warsaw Pact.

Lockheed followed the U-2 with an even more extraordinary family, popularly known as the Blackbirds. These were designed to fly not only very high but also very fast, and the entire design meant breaking new ground in structure, materials, engines, systems and even the ways missions were organized. The chief material selected was titanium B-120 alloy, which at first was found almost impossible to cut, shape or drill. Special fuel was developed, called JP-7, which in turn required special KC-135Q tankers of the 100th Air Refueling Wing. Even the hydraulic fluid had to be specially created; one fluid worked beautifully – except that if it was allowed to cool to room temperature it turned into a white powder! It is difficult to believe that the first of these amazing aircraft, an A-12, first flew as long ago as April 26, 1962. It was followed by an interceptor version, the YF-12A, whose radar and missiles still do not quite have an equal. Finally came the real production airplane, the giant SR-71 reconnaissance aircraft. One of them, after 10 years of operational flying, was taken without modification and used to set a world speed record of 2,193mph. Others, on the same occasion, were used to set a speed record around a 1,000km circuit and for sustained altitude, the latter figure being 85,069ft. Sadly, for financial reasons, the SR-71s were removed from the active USAF inventory in 1989.

Inflation is only one of the reasons why today we build slower airplanes. Whereas in 1960 numerous organizations had spent up to 10 years studying SSTs (supersonic transports) and in the 1960s various SSTs actually went ahead, today no-one seems concerned that the seven Concorde SSTs actually in service will before long reach the end of their working lives and that they have no successors.

An SST is very difficult to design. Its basic and almost crippling drawback is that the overall ratio of lift to drag, which may be 25 for a big subsonic jetliner, is likely to be only about seven for an SST (today designers are thinking in terms of 10, but that is still unimpressive). Because the airframe gets hot, aluminum alloys cannot be used but must be replaced by the much more expensive titanium or stainless steel. Instead of a relatively simple fixed-geometry engine pod, the engines need extremely complicated afterburner nozzles whose profile and area can be varied greatly (even when almost white-hot), and inlets of extraordinary complexity which have not only variable profile and area but also numerous auxiliary inlets and overboard dump valves and spill doors, all under computer control. The fuselage 'fineness ratio', which can loosely be translated as slenderness, has to be far higher than that of a subsonic transport. Thus, its interior either has to be extremely cramped or the length has to be impossibly great. And there are many other problems, not least of which are loud noise at airports and the sonic boom heard on the ground under the flight path. The latter, of course, was the crucial factor in killing off most of the SSTs and preventing the construction of new ones.

The first SST to fly was the Soviet Union's Tu-144 on December 31, 1968. This, however, had a hesitant and brief career in service, in contrast to the rather smaller Anglo-French Concorde. This first flew as a prototype on March 2, 1969 and was subsequently developed as a longer and heavier production aircraft, which entered service with Air France and British Airways on January 21, 1976. At one time it was hoped 200

***Right*: The Northrop B-2 stealth bomber is the most expensive airplane ever built.**

Rockwell is one of three companies working on the X-30 National Aerospace Plane passenger airliner concept.

Concordes would be sold, but it was almost killed by the steep rise in oil prices from the fall of 1973 and the fact that sonic-boom protesters succeeded in prohibiting supersonic flight over land. In service Concordes have proved popular and, considering their complexity, remarkably reliable. Sir Frank Whittle is one of the visionaries who has pointed out that, using the technology of 30 years later, we can and should be working on an SST for the 21st century. In France Aérospatiale has displayed beautiful models of two proposed SSTs, one a direct replacement for Concorde carrying 200 passengers at just over Mach 2 for use from about 2002, and the other a more challenging aircraft to carry 150 passengers at Mach 5 (3,300mph) for service from 2015. The Mach 5 aircraft looks, and in fact is, to some degree like a spacecraft. It would have quite a small delta wing at the rear of a huge flattened fuselage, which in cruising flight would provide most of the lift. As it would fly the whole of each journey within the atmosphere, even though at such a high altitude that the air would be very thin, it has to be very streamlined.

Spacecraft proper, however, need not be streamlined but need the structural strength, and a covering of refractory or ablative materials, to withstand the terrible stresses and temperatures of re-entry at about Mach 25 (16,550mph). So far the United States and Soviet Union have agreed on the same shape for their reusable orbiters – the American Shuttle and the Soviet Buran (Snowstorm). The main drawback of both, and of the Franco-German Hermès, which is to be a smaller vehicle of the same type, is that they have to be launched vertically by a giant rocket. We have only to imagine the problems and costs of using such a launch technique for every 747 flight to appreciate the disadvantages. The only good feature of these vehicles is that at least the vehicle itself is reusable, because it can be refurbished after making a normal

airplane-type landing. In the case of the U.S. Shuttle, the two SRBs (solid-rocket boosters), each imparting a thrust of 3,300,000lb at lift-off, are also retrieved after their descent into the ocean by parachute. But obviously the ideal would be a spacecraft that could take off like an airplane.

Ten years ago British Aerospace and Rolls-Royce began to develop the basic design of just such a vehicle, called HOTOL (HOrizontal Take Off and Landing). Not only would this take off conventionally from an ordinary runway, but the RB.545 propulsion system would start off using oxygen from the atmosphere, just as in other aircraft. This would make a great difference to the take-off weight and the volume of the fuel tanks needed. After about nine minutes, at a height of 85,000ft at Mach 5 (3,300mph), the engines would be progressively switched over to use the on-board liquid oxygen supply, their air inlets then being closed and the HOTOL becoming a true spacecraft running on a high-energy oxygen/hydrogen rocket system. It could, for example, place a payload of 17,635lb into a 162-mile equatorial orbit. On return the vehicle would weigh only about 104,700lb, less than one-fifth of its take-off weight, and it would land on a lightweight landing gear in a run of about 3,800ft.

HOTOL is obviously the way to go. Space travel will never be economic until people start using ordinary vehicles which throw nothing away and, whenever possible, use the oxygen freely available in the atmosphere. But HOTOL has the disadvantage of being British, so it has had to be developed on the proverbial shoe-string. In July 1988 the British government announced that it would leave funding to the private sector. But then the Ministry of Defense instituted Catch-22 by declaring that the brilliant RB.545 engine technology is secret, thus ensuring that the private sector cannot possibly participate. In due course, no doubt, some other nation will reinvent the HOTOL and it will succeed triumphantly. Still, Britain may then be able to rent or lease a couple of them – just to keep up with the competition.

First flown in 1984, the Grumman X-29A is totally unstable, and needs reliable computers to keep it on course.

Before undertaking its first voyage into space the U.S. Shuttle Orbiter was tested as a glider after being taken aloft riding on a modified Boeing 747 in 1977.
Inset: The similar Soviet 'Buran' Orbiter rides on the huge An-225 transport on its delivery flight to Baikonur.

NASA
905

A joint effort by British Aerospace and Rolls-Royce, HOTOL (HOrizontal Take Off and Landing) appears to be by far the most effective low-cost way to travel into or through space. It would use ordinary airports and be totally reusable.

An artist's impression of the first launch of the Hermes manned spacecraft from the French space center in Guyana planned for April 1988. Hermes is not totally reusable because the Ariane 5 booster is thrown away after each mission.

INDEX

Figures in **bold** *refer to an illustration*

ACKNOWLEDGEMENTS

The picture on page 10 is reproduced by courtesy of the Trustees of the British Museum.
Airbus Industrie 82-83; British Aerospace (Space Systems) Ltd 138-139; British Aerospace Commercial Aircraft Ltd, Airlines Division 98-99; British Aerospace Military Aircraft Division 2-3, 60-61, 62-63; Central Office of Information 54-55; Mary Evans Picture Library 6, 8-9, 12 inset, 20-21, 24-25, 26-27, 27; James Gilbert 112, 114 top; Bill Gunston 59; M J Hooks 9 top, 9 bottom, 31 bottom, 32, 38-39, 48-49, 53, 64, 70-71, 86 bottom, 88-89, 90-91, 95, 95 inset, 96 top, 104-105, 108-109, 116 inset, 117 inset, 118, 119, 120-121; Imperial War Museum 34-35; Philip Jarrett 85; Lockheed Corporation 77 left inset, 77 right inset; McDonnell Douglas 96 bottom; Novosti Press Agency 86 top; Quadrant Picture Library 68-69, 80-81, 136 inset; Quadrant Picture Library/Flight 31 top, 46 bottom; Science Museum 8, 12-13, 14-15, 18-19; SNECMA 119 inset; Frank Spooner Pictures 7, 73, 102-103, 113, 122-123, 128-129, 131; Taylor Photo Library 42-43, 52, 125; Tetra Associates 11; TRH Pictures 1, 28, 33, 34 inset, 106-107; TRH Pictures/Aerospatiale 140, /Boeing 100-101, /Department of Defence 44-45, /General Dynamics 50-51, /Grumman 66/67, 134-135, /Hughes Aircraft 92-93, /Alan Landau 74, /MBB 110-111, /McDonnell Douglas 4-5, /NASA 116-117, 136-137, /NASM 17, 46 top, /RAF Museum 36-37, /Rockwell 132-133, /SIRPA-AIR 64-65, /United States Air Force 74-75, 78-79, 79, 126-127, /United States Army 22, 56-57, /Vickers 114 bottom; United States Air Force 1, 76-77.

Book design: by Hussain R. Mohamed.